T0258080

IEE MANAGEMENT OF TECHNOLOGY SERIES 18

Series Editors: G. A. Montgomerie
B. C. Twiss

DESIGNING BUSINESSES

How to develop and lead a high technology company

Other volumes in this series:

DESIGNING BUSINESSES

How to develop and lead a high technology company

George Young

The Institution of Electrical Engineers

Published by: The Institution of Electrical Engineers, London,
United Kingdom

© 1997: The Institution of Electrical Engineers

The Institution of Electrical Engineers,
Michael Faraday House,
Six Hills Way, Stevenage,
Herts. SG1 2AY, United Kingdom

British Library Cataloguing in Publication Data

A CIP catalogue record for this book
is available from the British Library

ISBN 0 85296 891 4

Printed in England by Short Run Press Ltd., Exeter

To Elizabeth

Contents

Section III: FINANCE

5 Financial concepts for the technical professional

Financial statements. Profit and loss statements, cash flow statements and balance sheets; the relevance of such statements to business design. Accounting policies. Using statements for business design. Illustrative financial statements. The software consultancy. The distribution business. The finance needs of a business. The natural growth rate of a business. Strategies to minimise the finance needs of a business. Debt funding. When debt funding is appropriate. Types of debt funding.

6 Building the equity base of the business

The soft start-up. Opportunities for bootstrapping the business, i.e. avoiding reliance on external funding sources. The business plan. Objectives of the business plan. Structure of the business plan. Introduction. The market to be addressed, discussion of competitors and basic strategies to gain a position in the market and maintain competitive advantage. Description of products and manufacturing processes. Financial projections. Summary resumés for key promoters. How to raise venture capital, and the issues involved. The stock market. Public stock markets and the issues involved in running a public company.

Section IV: DEVELOPING THE BUSINESS

7 Alliances

The need for alliances. Examples such as Microsoft/IBM. Distribution and licence arrangements. Types of alliances.

8 Acquisitions

Why do acquisitions? Types of acquisitions. Relevance to the technical entrepreneur.

Section V: MANAGEMENT BUY-INS AND BUYOUTS

9 Management buy-ins and buyouts

The opportunity associated with buy-ins and buyouts. The challenges associated with such operations. Financial and operating issues involved. Timescales.

Section VI: CONCLUSIONS

Author's preface

Our world is getting more complex, with competition in many industries taking on a global dimension. To compete in this environment we need a broad skills base in our technical professionals, complementing a necessary excellence in core engineering activity. This is particularly the case when entrepreneurial activity is being envisaged.

Much work has now to be done through cross-disciplinary design teams, not just in bringing together teams of technical professionals from various areas, but also in including experts in marketing, manufacturing, finance and other key areas. The technical professional can no longer be isolated, designing to a very narrow specification.

This books seeks to address some of these broadening issues, with particular emphasis on entrepreneurial activity for engineers, and seeks to communicate points based on direct experience of the key issues involved in these areas.

The author would like to thank his colleagues at DCC plc, specifically chief executive Jim Flavin, for providing an environment where these thoughts could be formulated and communicated. Opinions offered are naturally those of the author, and responsibility for any errors and omissions rests with the author.

Section I
Introduction

Chapter 1
Engineers and business

This book is motivated by a desire to communicate the essentials of business strategy, with financial and related topics, to technical professionals. Specifically, the intended audience comprises engineers and other technical specialists of various disciplines who are

- contemplating starting their own business
- interested in undertaking a management buyout or buy-in in the context of an existing business
- finding that their role merits the description of 'corporate entrepreneur', leading a new business unit forward and exposed to many of the challenges of the entrepreneurial venture but without the direct risk and upside associated with owning the venture
- finding that they are increasingly working in teams comprising executives from a number of functional areas
- interested in becoming more familiar with the commercial aspects of the organisation with which they are associated.

Engineers are usually well placed to understand business issues. Their education and professional training emphasise a strong quantitative approach, based largely on the adage that if one cannot quantify an issue then it is not fully understood. Much of the required skill set for business also includes the ability to deal with quantitative issues, primarily those relating to finance.

Formal problem solving, specifically in design, is at the heart of the engineering profession. Design of an engineering system, with the necessity of taking into account all the variables and external factors impinging on that system, is analogous to designing an overall business. Engineers are continually being given wider design briefs, taking into account numerous factors in addition to the narrow technical performance. There are design for manufacturability (DFM) and design for test (DFT) concepts, and much product development is being done in teams, taking in financial and marketing considerations as well as technical input. Much design work now has to consider the complete life cycle of the product, and how it fits in in a commercial sense. In some ways, the design of a business is the ultimate extension of good product design.

Engineering communications skills also cover many of the areas needed in the context of business leadership, which increasingly is based on clear communications with various teams throughout the organisation.

The engineer is well placed to apply design skills and disciplines in the broader commercial environment in order to come up with the best commercial, as well as technical, answer. Some elements of the mapping involved are shown in Figure 1.1.

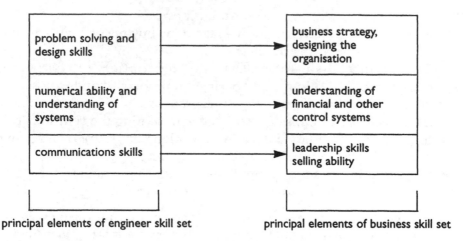

principal elements of engineer skill set   principal elements of business skill set

Figure 1.1   Mapping between engineer skill set and commercial skill set

In this book the fundamental approach is to build on the design skills possessed by the engineer and complement and broaden these skills to allow the reader to be better placed to start a business, to consider participation in a management buyout or buy-in opportunity, to work more effectively in broadly-based teams or just to broaden the vision of those engaged in product design work.

The link between engineering and business is acknowledged in such terms as re-engineering (currently widely used in a business context) and financial engineering. Re-engineering is applied to the redesign of many business processes to make them more attuned to customer need. The term financial engineering usually refers to efforts to design complex financial products to meet customer needs, or to designing the complex transaction funding structure which might, for example, be involved in a management buyout. A systems dynamics approach has also been used in the context of applying modelling techniques, usually initially developed for engineering applications in the modelling of physical systems, to the modelling of businesses and other social systems (KEOUGH and DOMAN, 1992).

Chapter 2 of this Section looks at why entrepreneurs start and develop businesses, and at the need for clear setting of objectives and for clear recognition of the opportunity and the nature of the task involved. One key aspect to point out is that one often has to look at the issues associated with exiting the business, i.e. realising value from it through a total or partial sale, *ab initio* if one is to do planning successfully.

Section II, on business strategy, then considers the key issues of designing a business in the marketplace, both in a top down sense looking at the business in the overall market context, and a bottom up approach starting from product design and working outwards to design many aspects of the business. The top down and bottom up approaches, as in the case of engineering design, are used together to come up with a business design which is self consistent and which takes account of commercial realities, being robust and capable of adapting to contingencies in the light of possible changes in the business environment. Also covered are the issues of technology policy, quality management, operations and recruitment and team building.

Financial considerations must underlie all business strategy. Financial issues are referred to in the business strategy section, but a dedicated section on finance, Section III, is appropriate. Here the basic financial statements are reviewed together with their use and value, and the finance needs of a business are considered, particularly with reference to funding growth. Options for raising money and the consequences of choosing each of those available are considered. Preparation of the business plan as a fundraising document is also discussed.

As the business develops it usually needs to develop alliances. Some of these will be quite informal, and some will be formal. Alliances can be very important, particularly to technology businesses, and, hence, a detailed treatment of this subject is presented in Section IV.

Acquisitions are also relevant to the entrepreneur, in that opportunities to acquire businesses can arise at a relatively early stage in business development, or the business itself may be the subject of an acquisition approach.

Finally, there are increasing opportunities for engineers and managers to participate in management buyout and buy-in opportunities. The corporate world is in constant change, with large groupings often deciding to concentrate on core activities, resulting in businesses not meeting certain criteria being offered for sale. Typically, businesses put up for sale are those that were acquired in the diversification programmes fashionable two decades ago. Alternatively, maybe an acquisition has gone wrong, or the acquired business was intended to provide a window into a new product area or geographical area. The experience gained may have suggested that the area is not really of long term interest, and thus the company providing this window is now for sale. Entrepreneurs who set up business in the 1940s and 1950s could now also be seeking realisation of their assets.

The first choice buyer is often the management, particularly if the alternative involves sale to a competitor. Selling to management also carries a good cachet of social responsibility, and can be good for continuity of the business, of interest to customers and of benefit to local interests in the region where the company is based. The selling entrepreneur may also have a strong desire to ensure that the business he has created finds a good home. Topics covered in earlier chapters are relevant to such situations, but there are specific issues, mainly associated with objectives of funders of such transactions, which deserve special treatment.

A key aspect of business relates to issues of commercial judgment, selling skills, negotiation and other interpersonal skills. These are the topics that are communicated least effectively in written form, but Chapter 10 outlines some of the important issues.

Business and entrepreneurship for engineers, and the related area of managing high technology companies, have naturally been very topical in recent years. (e.g. TWISS, 1988; BELL and McNAMARA, 1991; RIGGS, 1983). This book seeks to present an innovative approach based on experience of working with entrepreneurs as a venture capital investor. (A graphic illustration of commercial and technical life in a fast-paced industry can be found in PROKESH, 1993.)

For the sake of brevity we use the male form of pronouns and other terms, but the female forms can, of course, also be understood.

Chapter 2
Objectives of the entrepreneur or management team

Starting and developing a business is usually undertaken in order to achieve the personal ambitions of the founder entrepreneur or team. Going successfully through the entrepreneurial process and creating a business of value is obviously a source of immense satisfaction, added to which is the large financial gain that can be realised from a successful venture.

Every business is unique, with its own set of objectives produced either totally by the entrepreneur, if it is a one person business, by a team if there is a quasi-democratic or collegial relationship amongst top management, or by agreement between management and outside equity holders. As often as not, the true objectives are left unstated but reflect a complex mix of attitudes to risk, desire for achievement, desire for power, desire for real financial gain and other attitudes which reflect the complexity that characterises human behaviour patterns.

The objectives will also change over time. Having achieved success in a business and realised a significant part of the value created, the entrepreneur may be less averse to risk and prepared to aim at higher goals. On the other hand, having nothing to lose may have produced a relaxed attitude to taking on risk in the early stages of a business, with an aversion to risk setting in when there is more to lose!

Some businesses are founded with very limited goals in mind. Many entrepreneurs are quite happy to set up a business in the form of a consultancy, as a local service operation, or possibly as a franchisee, with the objective of providing an income of the same order of magnitude as could be gained in employment.

Others will have, from the start, the ambition of building a very large business, and will have the ability to convince backers to commit funding to the idea. Some businesses need very significant funding to start operations, and in these areas the entrepreneur has to think big from the beginning.

Entrepreneurs will also have different attitudes to risk. Some may be prepared to borrow as much as banks are willing to lend, giving personal guarantees quite readily. Others will be much more cautious, preferring to build the business slowly and apparently prudently. In some industries, however, one does not have the luxury of being able to proceed cautiously, and an apparently cautious course may actually turn out to be the more risky.

Some entrepreneurs will risk the company on the development of one project, while others will have a series of contingency plans worked out. Even large companies have to occasionally invest heavily in a sigle project. IBM's development of the 360-series mainframes in the 1960s is one such example, and indeed any major aircraft programme at, for example, Boeing, is really a critical decision. The reality is often that the main project is so important to *any* company that developing contingency plans diverts resources from the main task, increasing risk of failure to both the project and the company.

Entrepreneurial ventures can be based on a single figure or a broadly based management team. It is very easy to understand why many entrepreneurs prefer the autocratic form. It is undeniably a source of satisfaction to have power without the need to share it with others, and a one man band can be highly efficient. The lone entrepreneur, typically autocratic with a paternalistic streak, can make decisions far more quickly and act in response to market changes more readily than a large corporation.

However, in its purest form the one man band approach can only last for a limited time. Growing the business means having to take in additional management, who need to be given some element of empowerment if they are to really serve the company and contribute their ideas. The lone entrepreneur also loses out when selling the business, as a purchaser will look ideally for a business which is not dependent on one individual, no matter how talented. Building up a management structure that can benefit from the insights, and the ability to focus, which the entrepreneur can bring, but which has its own independent capabilities, is usually essential if one is to create value in the longer term.

Another aspect of power sharing comes from the desirability in many cases to take in outside investors. Some businesses can be started by using the founders' own resources, coupled with perhaps an unhealthy reliance on bank borrowings. In many other cases, however, the business needs to grow much more quickly than is possible using internally generated funds, or indeed it needs to start off at a certain minimum size if it is to succeed in its chosen industry. In these cases selling part of the venture to an outside investor, either to a financial partner or to a trade partner, or going public on a stockmarket, may be the only practical route forward. There is the saying that 'it is better to have 60 % of something rather than 100 % of nothing', which is often quoted by investors to the entrepreneur who may be reluctant to sell a share of the business!

Outside investors, however, have a fiduciary responsibility to their own funders. This will imply a need to appear prudent in taking decisions, which may impact on the style of decision making within the business. Usually, an external investor will come into a business on the basis of a clearly articulated development plan. If market conditions dictate that this has to be torn up and the activity of the business focused in a different direction, the outside investor may take some convincing. Some entrepreneurs can easily make the cultural adjustment to this broadening of the coalition of interests associated with taking in an outside private investor or going to the stockmarket; for others it is more difficult.

There is also the issue of personal commitment to a venture. Getting a business operation started can require an immense amount of straightforward hard work, sometimes recognised as valuable by investors and termed 'sweat equity'. As the business develops, a relatively stable company in an undemanding commercial area may give a very good lifestyle to the owners without huge effort. On the other hand, if the business is to keep developing, then continued periods of eighteen hour days may be required, even with fairly effective delegation.

The final issue is for the entrepreneur or management team to have some form of exit strategy in mind. Certainly, if outside investors join in the venture then some opportunity to realise their holding will be usually required, typically within five to seven years from the date of investment. This realisation will come most commonly from sale of the business, from a stockmarket listing which allows some share liquidity, or possibly from a refinancing of the company effectively to replace the interest held by an investor with investment from a new source.

For the entrepreneur or management team, financial objectives could well be met by having an ongoing profitable private company paying out generous dividends as well as good salaries to all concerned. In other cases, looking to float the company on a stockmarket can be attractive, as those who wish to sell shares can do so while others can retain their interest in the business. Selling the business in total, usually to a trade buyer, can also offer an appropriate exit mechanism if required.

For some entrepreneurs, exit strategies may be less relevant, particularly if the company is planned as a family business, maintaining continuity over generations.

The first step in design is usually one of establishing the specifications and the ground rules. In designing businesses, the desires and constraints imposed by the entrepreneur and/or the management team will be those which determine how the business is designed.

Section II
Business strategy

Chapter 3
Strategic business design

3.1 Evaluating the marketplace—the top down approach to business design

We thus come to designing businesses. The approach to be adopted has to be practical. One cannot afford 'paralysis by analysis', and must take note of the observation that 'by the time an opportunity is investigated fully, it may no longer exist' (BHIDE, 1994). A significant amount of resource, intellectual, emotional and financial, will be committed to the venture by the founder and perhaps by others. There is thus a duty, not least to oneself, at least to acknowledge and understand some of the issues involved. The real world requires operation with limited information in many areas, but knowing the key issues allows the risks we take to be at least understood if difficult to quantify exactly.

A significant amount of literature about both design techniques and methodologies for complex engineering systems has been developed (e.g. JOHNS, 1970; ASIMOW, 1963; WASSERMANN and FREEMAN, 1976). This ranges from approaches that are basically convergent in nature, seeking to undertake a thorough search for the best approaches to a particular problem, to divergent methods which stress the opportunity of achieving a very efficient and elegant design in those situations where effective use of innovation allows breakthroughs to be made, looking for ways of getting around the assumed constraints. Figure 3.1 summarises engineering design techniques.

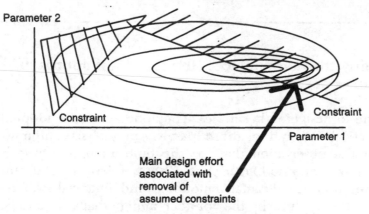

Convergent approaches - looking for the optimum solution within established constraints

Figure 3.1   Design techniques: a summary

Without seeking to get into a comparison of the various techniques, the basic contention is that engineers are well versed (formally or informally) in design methods, and emphasise thoroughness of evaluation of alternatives, attention to specifications and meeting these, and the inherent search for the optimum which characterises many technical professionals.

When designing a complex system, the basic approaches are usually as follows:

- a top down approach, where an overall view of the situation is taken, looking at initial designs without consideration of the associated details
- a bottom up approach where currently available or new technologies and capabilities are examined and how these might be combined to give the overall solution considered.

In practice, a combination of these techniques, and perhaps other approaches, can be used to select an optimal design which meets all specifications, or which gives the best compromise between conflicting requirements.

Designing a business is usually a more complex task than that associated with designing a product. In particular

- objectives and specifications are less clearly stated; objectives of entrepreneurs and managers are usually implicit rather than explicit
- the environment is in constant change, with ongoing moves by potential competitors and emerging technologies offering opportunities and presenting threats. These issues can be quantified very imprecisely, and yet important decisions need to be made which affect many people very fundamentally. This can be the hardest issue for those used to having to make decisions based on harder evidence than is often available in the business environment
- the twin tasks of designing the strategy to be followed and designing the fabric of the business consistent with executing that strategy must be performed. There is also a constant trade-off between keeping options open, i.e. flexibility, which has a significant cost in itself, and making a particular decision and having to live with the consequences. One is really designing a game playing entity, and this entity needs to be imbued with the ability to be innovative, reactive and learning (GARVIN, 1993)
- there is no unique business design, as may be the case with a product design. In designing businesses various alternatives are continually examined, seeking to extract the best result within vision at any given moment.

In the business environment, the initial business design approach may be top down. A business that has a clear role in the marketplace and which will be around for some time is required. The top down approach is concerned very much with undertaking an industry analysis, at least in sufficient depth to be able to see the main factors involved in building a successful business, and seeing how the venture will fit within that industry as it evolves.

Technologists are frequently in a situation where they have a product innovation to exploit. Not all innovations are market led— indeed, in the case of new products, the market may not know what it wants, and technology led opportunities, or just good ideas put into practice, are more common than some will admit. With a new technology it is important to see what products might be designed with it, what production processes might be compatible, what sales organisations and distribution channels are appropriate, etc. This is a bottom up approach to business design.

As in the case of engineering design, both approaches are then used alternately, searching for inconsistencies and seeking to overcome constraints, until the design has been refined to be the best possible given time and other resources available.

3.2 The top down design approach

In taking this approach, it is necessary to understand very well the type of market to be addressed. Every business operates in a competitive, or potentially competitive, environment and a company needs to know what will be required if it is to be active in this business area success-fully for an extended period.

Some examples indicate the key issues involved.

3.2.1 Looking at setting up a semiconductor business

A group of engineers believes that the semiconductor market is growing very rapidly. Its members have significant expertise in the area of DRAM devices and are looking at DRAM device manufacture.

Perhaps the first issue to be faced here is that DRAM (dynamic random-access memory) capacity comes only in very large sizes. There is effectively a minimum economic scale associated with a DRAM plant, with likely commitment of the order of £1 billion being required (mid 1990s).

DRAMs also represent something akin to a commodity, with significant price swings. They have been very profitable for an extended period during the mid-1990s, with demand led by the personal computer boom, but, given the large capacity increments which tend to occur in the cycle, historically they have exhibited some of the price volatility which one sees in other capital-intensive businesses such as paper processing.

What would it therefore take for the design team to get involved in this business? First, a very deep-pocketed partner with the ability to fund this level of investment. The venture capital community is likely to find such an amount of money quite daunting. These sorts of sums have been committed in the past by Japanese steelmakers and by Korean Chaebols to enter the business, but this group might realise that its members would effectively end up being employees of the backer, rather than leaders of their own venture, given the amount of funding required. This might be acceptable, particularly with good employment and incentive arrangements, but possibly not.

Thus the search to reshape the venture. Looking at the top down design approach might suggest that the group should look at other areas where barriers to entry are less daunting, and where it might be better placed to win business.

Operating a design bureau for devices could, for example, be a better option. With continually improving product and process technology, the group might be able to consider supplying design services to chip makers, possibly on a shared savings basis.

3.2.2 Considering a specialised assembly business

A team looking at setting up a business operating in the manufacture of specialised assemblies.

This industry is obviously less concentrated than the DRAM market, but looking at other firms within the sector we might see that key characteristics are as follows

- size—companies above a certain breakpoint in annual revenues look to be making reasonable returns, although companies below this figure are poor performers
- profitability of firms meeting the size criteria indicate a return on capital employed of 20 %
- quality standards. Approval to ISO9002 seems mandatory
- the business does not appear to be unduly recession sensitive, with volume growth predicted at 5 % to 15 % annually in the medium term
- a good, stable customer base of companies seems desirable; most suppliers in the industry become quite dependent on one customer over time, and this limits flexibility.

The industry that the team is considering entering looks moderately attractive at best. Inadequate returns are likely until a certain size is reached and industry competition will increase, as existing firms can cope with growth. Being tied to one customer is also likely to limit sales opportunities, and thus exit valuation.

Such an analysis reveals other ways of achieving objectives. Are there ways of adding process technology to stand out above other firms and get a real competitive edge? Could the team do a management buy-in to an existing underperforming firm, thus getting scale from the start? Would backers provide funding for going into the industry in this fashion, or could soft terms be organised with the vendor? Is there a 'consolidation play' possible in the industry, where most large customers are growing weary of the industry fragmentation and would prefer to deal with better managed businesses with multiple facilities? Can the management capability level in the business be raised and used as an area of competitive advantage?

3.2.3 Setting up a specialised engineering software business

Our team is looking at launching a software design business, developing products for design of niche telecommunications systems.

An industry analysis here might suggest that the UK market for the products is £1 million annually, growing at a 30 % annual rate, with this rate looking quite sustainable in the medium term. The outlook suggests that the market might be worth £3.7 million annually after five years. Customers include major wireline and radio-based telecommunications operators, and suppliers of test equipment to these operating companies.

Internationally, the market could be around 20 times this figure, reflecting the fact that the UK market constitutes about 5 % of world market for specialised high technology products. After five years, the remaining EU market is likely to be about £15 million and the US market about £40 million.

There are, perhaps, two other companies internationally at about the same stage in terms of product development.

Any top down analysis in this context would suggest

- with low international differentiation in products a company must compete internationally from day one. This would involve a world culture for the organisation, with much travelling required of founders
- such an industry is quite small in the eyes of major companies. It is likely that there will be four companies worldwide addressing this market after five years, perhaps two independent and two subsidiaries of larger engineering specialist software houses. It is a small field and the size of the industry is unlikely to be able to support more than four firms, given the levels of research and development and sales coverage needed. The new company must therefore be one of the four
- getting investment funds from venture capital sources may be difficult given the size of the opportunity, and perceived limited benefits.

If the team can succeed in this approach, well and good. Otherwise it might get a better return on product development effort by undertaking a royalty arrangement or other technology acquisition structure with a company which may be more suitably placed to be a leader in the industry.

3.3 The key issues in industry analysis

Modern approaches to industry analysis owe much to the work of Michael Porter, as communicated in his books *Competitive analysis* and *Competitive strategy* (PORTER, 1985; PORTER, 1980).

Businesses operate in a competitive environment. Rivalry between firms in an industry exists alongside pressure from customers (looking for lower prices, better performance, better delivery and longer credit terms etc), from suppliers (trying to raise prices, shorten credit terms, reduce specifications etc.), from potential new entrants and pressure from governments and other influencers.

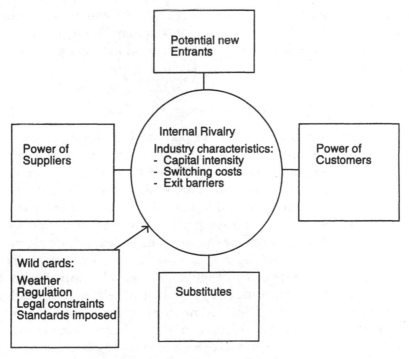

Figure 3.2   The key strategic determinants of industry profitability (amended from PORTER, 1980)

From the viewpoint of the business designer, an industry analysis is undertaken in order to determine whether there will be room for the business within the industry as it evolves, whether the industry will allow it to be profitable and what broad strategies should be pursued in order to ensure that the business is best placed. Also central is some understanding of how a competitive advantage will be maintained over time. Most technology companies start life with some clear product or process advantage, and maintaining this advantage, or deriving other areas of competitive advantage over time, is the key challenge.

It is very difficult to identify industries that are both good and automatically attractive. Indeed, the very perception that an industry offers lucrative opportunities can attract too many participants, with the result that it becomes too crowded and profitability falls. This is frequently the case with many technology industries. The overall sector may be experiencing excellent growth, but the wide availability of this knowledge can attract too many competitors, resulting in battles for market share and in poorer profitability than would obtain in a much less 'trendy' industry. Technology industries can also be characterised by extended periods of low profitability for firms within them, as they pursue some vision of Nirvana in which competitive pressure may ease in the future. And Nirvana may or may not arrive!

Generally, characteristics associated with good industry profitability are as follows.

Contained rivalry within the industry

This can occur when firms know their place within an industry, where that industry is segmented into various areas with an acknowledged specialist in each segment and limited competition across segments. A large company can maintain a price umbrella for smaller firms, and higher cost firms can be reluctant to cut prices, thus leaving low cost firms in the industry with very good profitability characteristics.

In some cases, competitors may be welcomed into an industry if it means that the incumbent will not have to develop extra capacity to serve marginal customers, or a competitor may be introduced by political and regulatory factors, e.g. in the case of telecommunications network service provision where competitive service providers have become the norm in most developed countries. In the semiconductor industry, and other industries where second sourcing is needed, there is a clear need for a competitor to be invited into the industry, which creates opportunity.

Moderate threat of substitutes

Minimal threat of substitutes can result in a profitable but complacent industry, which does little about looking after its cost base. The industry can thus be very exposed when cheaper alternatives come along. In technology businesses, complacency regarding product substitution can be dangerous, with legions of companies seeking to provide substitute products for those currently available. For example, in 1996, the established personal computer architecture has been challenged by the introduction of low cost network computers, dedicated to deriving maximum benefit from the Internet with limited local computing capability.

Power relative to suppliers and customers

If the industry holds a key resource in short supply, then it is likely to hold greater bargaining power over its customers than if a commodity in ample supply is being provided. Industries can go from shortage to glut very quickly, with consequent changes in pricing and in power shifts. DRAM components have, for example, been alternately in glut and in scarcity during the 1990s. If one can get into a position of being the only or dominant supplier of industry standard products, then obviously the power balance relative to customers is strong. Such a position explains the very satisfactory margins enjoyed by Microsoft on microcomputer software and by Intel on processor devices.

Having a strong patent position can also be attractive, as a patent gives one a state granted monopoly for up to twenty years on the relevant technology. And monopolies are very attractive to their owners. Commercial and legal constraints usually prevent full and abusive exercise of monopoly powers, but even a quasi-monopoly is better than a fully competitive environment.

High entry barriers and low exit barriers

An industry is obviously attractive if it is not too crowded, and this is aided by the existence of entry barriers, which make it difficult for a new entrant to gain a significant position. Entry barriers can be scale related—a larger firm can spread the cost of research and development over a much large customer base than is possible in the case of a smaller firm. Also, size may well be correlated with low cost of product and with having worldwide distribution capability.

Sometimes overlooked are issues relating to exit barriers. Basically, it helps if companies that want to leave an industry can do so easily. If they stay unwillingly in the industry, they can price close to marginal cost and keep overcapacity in place, thus damaging the business environment for all participants. Exit barriers may come from having large amounts of dedicated capital equipment, an inability to close a plant based on social or government factors, etc.

If one is looking to enter an industry, one has to form some views as to how the industry will develop over time. Some key characteristics to understand must be the following.

Industry concentration and scale issues

Many technology based industries concentrate over time. The industry follows the classical life cycle approach (Figure 3.3) where in the early stages barriers to entry into the industry are low, and customers are excited about a product which takes perhaps three man years to develop. Five years later, the industry may have grown very substantially, and growth rates may still be high, but the typical serious product now takes 100 man years to develop, and worldwide distribution and marketing capabilities are necessary if one is to be taken seriously. The area of microcomputer software has seen, for example, development from the initial Visi-Calc product, through to the initial Lotus 1-2-3 product to 1990s standard complex spreadsheets with graphing capabilities and multiple levels. Each generation of product represents an order of magnitude in complexity increase as the industry progresses, and if a company cannot stay up with the development path, then it is, literally, lost, with the business having almost no value.

The industry may be quite attractive after some time, when the battles are won and the survivors can enjoy an environment of contained growth, with high entry barriers making it difficult for newcomers to enter the market. It is rarely possible, however, to rest on one's laurels in technology businesses! (The life cycle concept has been broadened into a fuller treatment of the ecology of development and survival in an industry, for example as presented in MOORE, 1993.)

There are naturally many industries that will allow companies across broad size ranges to make money for an extended period. Many local service businesses fall into this category. One needs, however, to be wary of situations where industries can consolidate quite quickly, and where participants can be faced with a sudden 'get bigger or get out' decision.

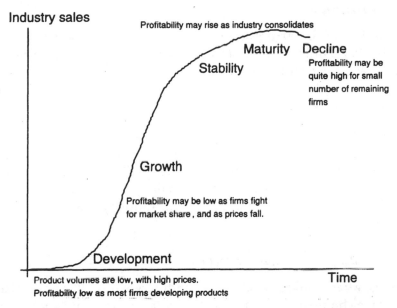

Figure 3.3 *Industry life-cycle*

Industry growth rates

In a growing industry there is usually more room for new players. Likewise, as an industry enters a downturn, customers may be more supportive of existing suppliers and the existing industry may be more likely to close ranks against a newcomer.

Receptiveness to innovation

An industry with a customer base which is receptive to new ways of doing things can give opportunities to a new entrant. The office supplies industry has, for example, been changed in recent years by the development of superstores, as a new class of entrepreneurial small businesses has been prepared to be innovative in its purchasing of office supplies.

Emerging standards

If a standards battle is raging in an industry, a company may be forced to take sides and risk becoming the loser. Standards battles rage from time to time within most industries, with the VHS/Betamax battle in video recording probably being the most famous. If a company chooses a computer platform or other solutions where it is effectively choosing a standard with which to be allied, it must be conscious of the need to pick the winner.

Alliance structures

Some industries break down into various camps, and being associated with the companies in one camp may make it difficult to do business in other areas within the industry.

Homogenous or niche

In terms of customer demand some industries may be quite uniform, while others may have different products and customer groups which become small industry areas in themselves—effectively niche markets. Areas such as instrumentation are usually very well populated with specialist niche markets while the overall industry has some common threads.

Understanding the key success factors

Knowing what it takes to win in an industry is usually vital. In some cases it may be mainly product performance, in some it may be having a superb service network, in other cases it may be in deployment of new customer accessible technology (such as having customers able to track parcels through logistics channels, or having customers check status of stock at the warehouse). Another key success factor is bundling. Microsoft, for example, bundled together all the major office applications of spreadsheet, word processing, presentation graphics and database as a suite. These suites then effectively became the only vehicle for mass market sale of these products, to the advantage of Microsoft, which had offerings in all areas, but to the initial disadvantage of Lotus and WordPerfect.

3.4 Strategies for entering a new industry

If a firm is entering a new industry, the focus has to be both on winning market share and establishing the standards. In effect, a race is being run between companies to gain access to key resources such as people, sites, money and the other elements needed for business, and key customers and partners must be signed up at as rapid a pace as is possible.

High market shares in the Internet area were initially secured in 1995 by Netscape Communications Corporation, with an estimated 85 % share of the browser market for viewing world wide web pages. To position Netscape Navigator as the industry standard, large numbers of copies of the product were given away to users, and made available for free downloading from the Internet itself. Establishing a product as the industry standard attracts investors at high prices—the valuation of Netscape peaked in late 1995 at $6.5 billion with reported quarterly revenues of $20 million—and can provide user endorsement, e.g. 'this page is best viewed using Netscape'.

Winning the race usually requires effort.

Technology development capability

The ability to make products that lead the industry in terms of performance, or that are indistinguishable in the minds of buyers from the leading performance products, is important. Getting the products established as the *de facto* industry standards is the key challenge. Essentially, in these days of industry norms, a company's products must be compatible with established industry standards while offering performance enhancements going beyond the basic industry requirements. The issue of architectural competition, where companies seek to get effectively proprietary positions in an ostensibly open environment, has been treated in MORRIS and FERGUSON, 1993.

Understanding the nature of the game and the size of the issues involved

This can be an area where companies in Europe, with more fragmented markets, can lose out when compared with companies in the US. If a company comes out with a new technology, there are very few borders that apply (unless there are different standards in various areas of the world, e.g. in some telecommunications businesses) and it is competing in the world market *ab initio*. The company may be able to find a backwater or niche application, but for a mass market software product or a product such as a disk drive or a personal computer, it is up against the best the world can offer. This means that the venture has to be resourced properly in order to take a meaningful level of international market share and make it worthwhile. If *de facto* standards have to be set, the task is greater but the longer term advantages are correspondingly greater if the company can achieve an industry standard solution.

Timing issues are also important. An unduly early entry wastes resources and invites disappointment, but a company can also just be too late. Some markets take a frustratingly long time to burst forth, while others advance quite rapidly. Selling industrial products can be initially quite slow, but then sales can accelerate as the new industry standard is established.

The futurist Paul Saffo has outlined the paradox that attends commercialisation of many products (SAFFO, 1985). Being able to see clearly a vision of what is possible does not necessarily mean that mass market acceptance of a product is imminent. He uses the term macromyopia for the situation where initial high hopes for an early version of a product get dashed. The product then typically languishes for a time before being taken up gradually, moving on, in later versions, to take on the mass market role as earlier envisaged. The personal computer, for example, was greeted in the early 1980s as being an ideal tool for the home, but it had to await the greater power and capabilities associated with mid-1990s models before home penetration levels became meaningful. In this light, getting the timing right is very important. Frequently one can have an elegant offering that is simply ahead of its time or, with luck and judgment, one can catch a very fast growth wave.

Getting resources

Building up the necessary market presence requires access to resources. The first of these is undoubtedly management; a capable management team is essential for competing with the best that the world can offer.

Closely allied to this is money. If a company is entering a new industry, it is important for it to tap resources while investment in the industry remains fashionable. After several experiences during the 1980s of the herd mentality, where investors committed money in excessive amounts into, say, the hard disk drive industry, venture investors are more astute and the venture capital industry is more likely to realise at the right time that enough money has gone into the industry in question. To take advantage of available money from investment sources, it may be necessary to move very quickly before supplies dry up. Failure to secure such funding leaves one poorly placed relative to well funded competitors.

A company seeking to build a significant business will need to secure appropriate distribution channels for its products.

Building market share and winning industry acceptance

A good customer list is probably the best calling card possessed by a developing company. It shows endorsement and it implies to the buyer or investor that he is at less risk than if such customer endorsements were not in place. Building market share with leading companies can require a lot of expensive development work and in some areas, selling the product cheaply or giving it away to win market share will win business. In other areas this approach may cheapen perceptions of the product and damage the company doing the promotion. Knowing exactly the buyer behaviour involved is a key business skill, and allows one to make the judgment involved.

It is important to be sure also that winning the race is worthwhile. There are many instances in business of Pyrrhic victories, where the battle has been won but the victor finds the resulting industry conditions to be unattractive. When fighting battles for market share in an emerging industry, and taking appropriate pain in terms of profitability, a company must be able to offer some vision of a future environment where competitive pressures can ease and the profitability rewards which result from having good market share in a more stable industry can be gained.

3.5 General strategies for entering, or winning further share in, existing markets

The head on approach

This is where a company competes head on with the incumbent; frequently the existing company is unable to respond, at least not in time to avoid loss of significant market share. The cost structure may be too high, technology may be dated and the understanding of buyer behaviour may be outdated. The existing company may also be wedded to stockholder expectations of high earnings and dividends, and may thus be unwilling to cut prices.

Finding such soft competition is less easy in the 1990s, as years of recession earlier in the decade, following on waves of restructuring in the 1980s, drove many of the more inefficient companies out of business. Finding the soft target cannot thus be taken for granted as easily as it might have been for new companies in the 1980s.

To take on the existing company in the head on sense requires some combination of demonstrably better technology or distribution skills, better understanding of buyer behaviour and better customer support. The counter attack must be unlikely to succeed and it must be certain that the firm losing market share is indeed constrained by high costs and an unwillingness to cut prices, and does not have a technological advantage about to be launched.

'Know thy enemy' is a well established adage, and it is important to know how a competitor can react. A partial list of options is as follows (PORTER, 1985)

- closing off niches by buying up companies which serve these niches
- using price fighting brands (particularly seen in the grocery industry) to deter entry to segments being considered by a newcomer
- taking firmer control of distribution channels, and bundling products
- raising switching costs through training and other linkages such as computer based ordering
- increasing scale economics by boosting advertising spend
- increasing capital expenditure requirements
- maintaining a presence in all technologies that could be relevant
- tying up supplier capability
- raising input costs for competitors.

A broad range of tactics is available to the incumbent to fend off the newcomer, from closing off potential alliance positions and supplies to direct persuasion of customers. The agile competitor will use such techniques to maintain market position against an unwanted entrant. Some of these techniques may well be construed as anti-competitive behaviour and the incumbent company may become liable to penalties, but this is of little comfort to the competitor whose business may well have collapsed in the interim. Having right on one's side is desirable, but it is not usually enough in such battles!

Competing indirectly

There may be occasions when direct head on competition can work well, and this may be the best strategy. However, there are other ways of achieving the same result, with less contention. These include:

Going after less favoured customers, or exploiting customer rivalry

This situation often exists in service industries, with the customer group dominated by a large customer, coexisting with a number of others. One option is to claim to some of the 'others' that may have been neglected by suppliers in favour of the dominant customer, and thus developing supply alliances on the basis of perceived priority of supply. Supporting some of the smaller, fast growing customers can allow a company to grow on the back of these other companies, getting all the cost advantages of increasing scale. On the other hand, if weaker customers are supported, then the approach may fail.

Going after niches

Here a company goes after underserved segments, or areas not served optimally by the large companies which may try to implement a 'one size fits all' approach to customer issues. It might, for example, develop a software product for a more esoteric operating system environment. As long as this system exists for some time, it may be the only serious provider of the appropriate software and have a near monopoly. This can allow the company to grow to a level where it can then break dependence on this niche and begin competing in the wider marketplace.

Getting welcomed into the industry

When serving an unattractive segment, or being forced into the industry as a result of regulatory pressure, the competitive pressures may be very limited. Likewise, if the industry is running near capacity and current companies are reluctant to add capacity in fear of another downturn, the initial opposition may be very muted.

3.6 Specific issues relevant to technology businesses

Technology comes into business strategy issues in a number of ways. Businesses usually base competition on generic strategies offering either cost leadership or differentiation opportunities across the whole industry, or focusing on specific segments.

Opportunities for cost savings are offered by technology in numerous areas, ranging from the use of more efficient manufacturing techniques in the production area, to better control of marketing and outbound logistics operations. It also allows additional opportunities for serving segments or for offering the differentiated product range demanded.

By technology companies, we usually mean ones where product technology is seen as the main tool for gaining competitive advantage. A bank may be a very astute and advanced user of technology in many of its delivery mechanisms, but would not normally be understood as a technology company. A company competing with others in manufacturing integrated circuits, pharmaceuticals, instruments etc., would generally be perceived as being a technology company.

Technology companies compete in markets having the same generic characteristics; still very important are issues of power, segmentation and competition. Additional factors, however, characterise industries relevant to technology companies.

Inexperience of management

Many companies in emerging industries are led by technologists. This is to be expected, as a fundamental knowledge of the technology is usually very important, at least in the early days, and some technologists make excellent managers. However, many of the technologists have been either reluctant or unable to make the transition to managing a technology business.

It is all too easy to think that, having designed the product, the business issues will fall into place. Things just do not happen this way. The task of managing a company requires just as much skill, perhaps more so, than was deployed in managing the product development function. An environment of general inexperience in the industry can, of course, create an opportunity for the well funded, well managed company to stand out amongst competitors and become the partner of choice for major customers.

The speed issue

Things can happen very quickly in technology led industries. A lost six months can mean a very dated product, the loss of opportunities to recruit experienced management and, perhaps, the closing of the window on availability of further funding from venture capital sources. The company may also be deprived of the opportunity to have its product become a *de facto* industry standard.

The nature of the races associated with technology companies can thus be very intense. Perhaps the most publicised technology race in 1995 was for continually improved Internet browser software, as companies sought both to incorporate some of the Java™ capabilities into their browsers and to use techniques that optimised performance on low bandwidth links, such as the use of progressive display of images.

The international nature of businesses

Technology is a worldwide phenomenon, with very little opportunity of keeping an advance local to a particular country or region. It can be necessary to go global very quickly with a product, and this involves committing resources, setting up alliances and other techniques for getting the best position in the overall market.

3.7 Forecasting

Implicit in any top down design is an understanding of the major trends at work in the industry. The well established life cycle concept may be observed over time, where industries move from rapid growth through to consolidation and maturity. The experience curve, where costs to producers fall with increasing volume, is also usually at work. In designing a business, it is necessary to plan for developments consistent with these phenomena.

A technique that can be of value here is scenario analysis, where a potential entrepreneur tries to come up with self-consistent views of how the industry will appear in, perhaps, five years' time. It is naturally a very difficult exercise, but going through the process at least forces one to think about how the shape of the industry will be viewed. Several scenarios can be examined, and perceptions of how the industry might evolve based on several sets of assumptions. Some of these sets of assumptions will give a very favourable environment for the proposed venture, while others will encourage examining alternative possibilities for coping with downside or pessimistic scenarios.

Entrepreneurs or managers are betting their future or career on a venture whose value will only be determined some years from now, so they must make some judgment as to the likely business environment. The successful entrepreneurs are usually those with good judgment, even (perhaps especially) in difficult situations; it is necessary to form some view of what the business environment will look like around the time when real value will be expected from the venture.

3.8 Quantifying objectives

The entrepreneur or the team will have objectives; if a backer is being taken on, then there will effectively be a coalition of objectives. Before setting forth, having a clear objective for the business in the context of its chosen marketplace is desirable, and this can then be checked for consistency in the light of resources available, profitability objectives, etc.

The target market might, for example, be seen as being worldwide worth £200 million after five years. Growth will be dramatic in years one, two and three, but is likely to have stabilised to 10 % annually in year five. There are no technical barriers internationally, and thus the world is considered to be the market. In markets of this character, having less than 20 % market share is unattractive, therefore the approach must be to get to £40 million in annual revenues within five years. The characteristics of the market addressed suggest that it will be quite profitable in year five, with profits of 20 % net pre-tax (with good return on capital) being a reasonable expectation, having allowed for 8 % continuing R&D (research and development) spend. To get to this position one needs

- extensive ongoing R&D commitment
- setting up of an international support network
- to win meaningful market share worldwide against locally based companies. The effort needed to achieve this must be costed, and if excessive, strategies involving alliances developed.

To get to the desired end point, effectively involves a race with a number of competitors. The company needs to be as agile and as able as any if it is to win market share and get endorsement from major users.

Business objectives must be expressed in terms of market position rather than in terms of absolute size or profitability in a given period. Size or profitability indicators reflect conditions at any given moment, but businesses with lasting value reflect success largely in their market position, this itself based on the strength of the products and services offered. Profitability is then derived from this market position, and from good internal management of the business.

Business is also about continuity. One must plan for the ongoing development of the business, taking into account R&D and market development costs.

3.9 Contingency strategies—designing for sale and limiting functions

It may well be that the vision of a full function business, taking in responsibility for product development, manufacture and worldwide sales, does not come together in a fashion consistent with the personal objectives of the entrepreneur or management team, or with the reality of being able to raise money. In this case concentrating on continued product development and on developing the local marketplace is an option. Although this is not a sensible long term approach, it will build value likely to appeal to potential acquirers as the industry consolidates. And industry consolidators are prepared to often pay high prices for market leadership in a local market, as for example indicated by transactions such as the 1995 acquisition of Unipalm/Pipex (the UK leader in the Internet service provision area) by UUNet of the US, later acquired by Worldcom.

Designing a business to be acquired makes sense if clear value can be shown, as evidenced by having good technology, high market share in a local/regional market or some other strong intangible assets. These could include a strong management team able to offer capabilities to the acquirer. In this last case the acquisition price is almost a transfer fee for the management team.

However, designing a business to be acquired has risks, as follows

- the number of potential acquirers may be quite small, and the hoped for auction not very competitive. If potential acquirers are able to go into the business area themselves, this can be held as a threat which can reduce value
- if rivals get acquired first then the value of the business may be significantly impaired
- choice of timing may be very important. Too early, and there is insufficient value attached to the business, and too late, most potential acquirers are already ensconced in the market, with consequent reduction in value.

Particularly in the pharmaceutical industry and medical device industry, there is a focus by large companies, such as Johnson and Johnson, Pfizer and others, on both acquiring promising developing companies and undertaking venture capital investments in up and coming businesses. In the electronic documentation area, Adobe in 1995 announced a venture fund to be managed by Hambrecht and Quist of San Francisco with a focus on technologies compatible with Adobe's development plans in the documentation area.

The venture backed company model may offer a more productive product development environment than can be achieved in corporate laboratories, or at least complement work done in such laboratories. A number of entrepreneurs in the healthcare field have taken the route of developing an innovative product, proving the market locally, building a good production facility and then presenting the multi-national healthcare company with the ideal acquisition. Dr Mario Veronesi of Mirandola is, for example, cited as having undertaken this approach on four occasions over a near twenty year span, each business being separate and non-competitive and having been sold to a different multinational acquirer.

If a business is not designed to be acquired, or if the plans do not work out in the intended fashion, there may be a role for the business as a contract manufacturer, as a design house or as a local service facility. These ambitions may correspond to a less grandiose vision of the world than may be have been initially in the eyes of the founders, but business is about reality. Concentrating on strengths in such areas may lead to businesses which are more profitable and less competitive and risky than full function product design/manufacturing/sale companies.

In some cases a hollowed out version of the original approach, where as many activities as possible are outsourced to others, could be considered. These activities might include production and logistics issues associated with the product, and possibly marketing and customer service. For example, changing patterns in the microcomputer software industry have brought about the creation of full service operations looking after all production and logistics operations, help line handling, translation/localisation and customer service, billing functions and cash collection. Being able to rely on such infrastructures may make the original plans more viable.

Opportunities for hollowing out the business in some functional areas are shown in Figure 3.4.

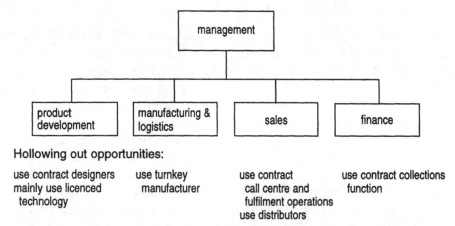

Figure 3.4   Outsourcing or 'hollowing out' opportunities by functional area

3.10 The top down business design approach—summary

In looking at a venture, one thus needs to understand a lot about the target industry and where the company is going to be positioned within it. This includes knowledge of how to play the game, and indeed what the rules of the game are. In many cases, if one is to recruit colleagues and take on backers, one needs to be able to communicate very clearly a knowledge of the industry and a game plan for addressing the industry in question.

3.11 The extended product design—the bottom up approach

The bottom up approach is focused on the business and on its immediate customer, with less initial emphasis on the overall industry structure and competitive issues. This, perhaps, represents the traditional approach to developing strategy. It remains a valuable tool in itself, but it is best complemented with the top down or industry analysis view as communicated in the previous section.

A bottom up approach can be regarded as an extension of product design and a good product designer will not stop at the design specification. When designing an amplifier, it is usually not enough just to design a printed circuit board with a few components inserted to give the desired gain, impedance, stability and frequency response characteristics. Consideration needs to be given to issues such as

- how to design for efficient production. Can the product be tested adequately? Are there enough easily accessible test points? How can automated functional test jigs be designed?
- the real cost of the product at various volumes. What are the commercial implications associated with use of singly sourced components? Is it worth paying more to use industry standard multiply sourced components? Is it better to reduce the number of component types, even though in some locations overspecification parts are used?
- the sale of the product. Does complementary equipment need to be designed for testing and demonstrating the product in the field? What should the warranty policy be? Should failed boards be fixed in the field, brought to base, or scrapped?

Following logically forward from any product design leads to designing the complete operations environment, taking in product and sales facilities, in effect designing many aspects of the business.

Figure 3.5, shows the design task logically expanding beyond the product area to take in production, marketing issues etc. This approach of integrated design, taking in marketing and production and effectively designing much of the business, is consistent with the

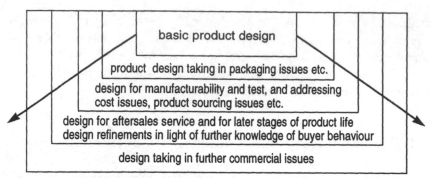

Figure 3.5   Broadening of the product design remit takes in broadening ranges of commercial issues

current trend of re-engineering, and includes tying together activities which may previously have been unduly fragmented and building over many of the turf barriers that characterise functionally fragmented organisations. To get speed into the organisation and cut time to market, it is necessary to employ this integrated, multidisciplinary approach to product development. The old days of marketing handing over a specification to engineering, which then gave it on to production, are gone. The weaknesses of this approach, all addressed by having a broader approach, include

- marketing not being able to understand the engineering trade-offs involved, and not understanding what may be possible with the newest technologies. It may be that of five features sought, the most trivial could be the most costly to provide, while other features could be provided at virtually no cost with some extra software changes
- design not really taking in ideas of product life cost, and not considering costs of manufacture, of service in the field, of use and finally of disposal. The critical role of design in influencing product cost has been outlined in COOPER, 1996 and the importance of design in a strategic sense has been treated in LORENZ,1994.
- losing out on opportunities to optimise production, based on making the product ready for easy manufacture and test, and giving the product as much self test capability as possible.

An idea for a new medical electronic instrument might start with the product design based on initial understanding of the market. The feature set is well defined and the product can be sold at a gross margin of 50 % at a sale price of £4000 to distributors.

end preconceptions:

- marketing needs to view engineering as creating value, and having a broader remit than just coming up with a sheaf of drawings (or the computerised equivalent)!
- engineering needs to see marketing as the ambassador of the customer within the company
- the customer needs to be understood in detail, and marketing must communicate the key buyer behaviour issues if the business is to succeed
- marketing and engineering functions are also part of the broader business; large investments often need to be made for new products, sales organisations have to be revamped etc
- marketing staff must be happy to let engineering staff talk to customers to assist in understanding buyer issues.

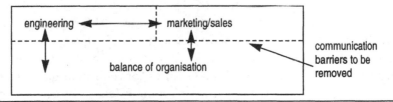

Figure 3.6   A need to break down barriers—marketing and engineering areas

Such an instrument represents a complex sale, in that the device must be qualified for reimbursement (or the practices performed by the device). The product must also find approval with doctors, nurses, patients and administrators and thus several sales tasks need to be undertaken. Understanding buyer behaviour is also central to profitability as it provides information about the price that the buyer is prepared to pay for various features.

objectively, the buyer may need a product, but buyers buy what they want, not necessarily what they need!

for many products, the purchase initiator may be difficult to identify, or there may be many buyers e.g. in selling a medical product one must address
- insurers (in order to ensure reimbursement)
- medical professionals, specialist and generalist
- government (for approvals and reimbursement issues)
- patients themselves, who may favour the product—e.g. patient demand has been an issue in the growth of minimally invasive surgical techniques

understanding the buyer behaviour is essential in so many commercial areas, ranging from product sale to alliances and acquisitions

Figure 3.7   Understanding buyer behaviour

Design of the product *per se* can then proceed smoothly, taking into account all appropriate safety norms and best practice ideas.

Continuing the design, likely production capacity—inferred from a market study—might be about 500 units annually. Production of the units could be subcontracted out with a final test capacity in-house. Doing this requires clear design of the information package for the subcontractor, unless staff are to be on hand at all hours to assist with production start-up.

In the sales area an alliance in the UK might be set up, and this market developed for the first year. For the first year after product launch the distributor can be deliberately oversupported to give extra enthusiasm about opportunities for picking up margin and most importantly for getting initial customer feedback. The design task takes in the development of appropriate communications material, in the form of brochures, manuals and other media, for all key specifier and buyer influencing groups. Release times for technical papers from medical researchers endorsing the product, release times for press comment about the device and the whole orchestration of the sales campaign should also be included in design.

Field service could be set up on the basis of a return to base system, with replacement units available at each regional office of the UK distributor. This should effectively allow four hour replacement of the unit in areas corresponding to 95 % of the UK population.

The approach is again one of broad design, allowing the design of the product to expand very logically until all key issues have been addressed in areas which go well beyond product design. By operating a broad canvas approach, features which might be incorporated in the product to improve usability can be considered. Improved self test capabilities, and perhaps a jig which could be located at each sales depot for evaluation of any defective units, allowing local board-swapping, might be attractive. Using remote diagnostics, possibly with a modem, is also a possibility.

Complementary products should be considered. Most products cannot exist in isolation—to sell cars, fuel must be available! This issue can, of course, be turned to advantage. Many instrument makers sell the product at a reduced margin in the expectation of having a long consumables trail, where users will buy high margin consumable products to go with the equipment. There is naturally an inertia that drives buyers to buy consumables which are warranted by the manufacturer as suitable for the instrument in question, and the manufacturer thus can usually charge a higher price than might otherwise be feasible.

A checklist for points to be addressed during this wide canvas design approach includes

(i)   Understanding buyer behaviour issues
    —knowing who the key buyers and influencers are, and what
       they need from the company. Are there committees involved,
       and is there a wide circle of influencers to be converted?
    —buyer budgetary allocations, timing, approvals thresholds
       and procedures
    —design of brochure material and other communications
       items
    —are there many types of buyers, i.e. is buyer segmentation an
       issue, and are different products needed for each segment?
    —what are the issues of direct sale versus using a distributor?
       Will a distributor take sufficient interest in a new product?
       How will distributor staff be trained and motivated?
    —can trade shows be used as the main means of
       communication?
    —how effective is direct mail?
    —does each sale need a demonstration?
    —cost of sales calls and conversion rates? The cost of selling
       the product can sometimes be just too high. A £5000
       product that requires a demonstration in order to get each
       sale, with most orders for unit quantities and a modest
       conversion rate from demonstration to order, may not be
       making money
    —leadtimes from initial contact to order, and on to payment?

(ii) Getting the complementary product issues right and ensuring
availability of consumables, considering also the profit
opportunity involved.

(iii) Looking at distribution, logistics and service
    —how does the product physically get to the customer?
    —can it be set up by the user?
    —are there training opportunities available?
    —is a nationwide maintenance organisation needed?
    —which sales functions are done in house, and which should
       be given to a distributor?

(iv) Production issues
    —to what extent do we 'make or buy' subassemblies?
    —should all manufacture be contracted out, or should some
       test functions be retained in house?

(v)   Getting the resources needed
    —is adequate manufacturing capability obtainable?
    —what staff training is required?

(vi)   Procurement issues
—can all components be bought?
—are components available from multiple sources or only
single vendors? Is this risk acceptable?

(vii) Product development issues
—is the company well placed for continued development?

Any real product design thus has to move very rapidly beyond the narrowly technical to take in many of these issues, and to ensure that it is fully in tune with business requirements.

3.12 Integrating the top down and bottom up approaches

Design of products is an iterative process, where results from one approach are weighed against constraints obtained from taking a different view of the design task.

Likewise in the case of design of businesses. Looking at the overall industry in the top down approach gives an idea of the likely scale needed, and gives some pointers for playing the competitive game. Coming from the bottom up approach allows various aspects associated with the business to be pulled together and the scope of strategies to be adopted for product specification, selling and production to be developed.

As an example of pulling together the different approaches, consider the design of a possible business to exploit an innovation in the area of semiconductor sensors. The innovation is focused on the detection of gases given off by deteriorating food products, and provides a quick red-amber-green indicator of value in, for example, meat and produce departments of supermarkets, in restaurants and in the home. *The example is based on an idea which might be feasible some years from now, but is intended to be fictitious with no representations of real firms involved.*

Initially there is the business definition issue: is this the semiconductor industry or the food technology industry? In this case both options can be considered, selling the sensor chip to instrument manufacturers or becoming a food instrument company.

Looking initially at the food instrument industry, this consists of a number of companies serving mainly laboratories with equipment which is quite costly. The particular niche of supplying equipment to retailers and restaurants to check food quality has not really developed, and will only develop when the law requires use of such technology. This observation has interesting consequences—first, the market will not develop until such laws are brought in, which might offer some staging of the market as laws come into effect at different times in each region or for each type of food outlet. Secondly, the market might be very large for a short period in the case of each country or each type of outlet as customers seek to comply with new laws, before settling down to being a replacement market. In a democracy, it is naturally the right of any interest group to lobby lawmakers and regulators, and the company might seek changes to legislation in such a way as to favour its interests.

The competitive threat might appear quite distant. Other semiconductor companies are getting sensors ready, but very few food instrument companies appear to be addressing the market. However, a significant threat may come from some of the larger food laboratory instrument companies which are looking at the emerging segment with some interest.

On the semiconductor side, meanwhile, progress is ongoing. It looks as if there will be three serious competitors here, another specialist company and two broadrange semiconductor companies with complete fabrication facilities. The judgment is that the integrated companies will be able to drive sensor cost down faster than the independent companies, including the proposed business, as it will have to rent processing capability. Two sources of processing have been identified, but pricing from these sources may not be such as to match the advantages held by the integrated companies.

Key issues associated with the emerging industry (defined as retail/restaurant electronic food condition monitors) are thus as follows:

- very fast growth likely as regulations are enacted, with minimal sales prior to this
- product very dependent on supplies of key sensor chip, with all other aspects of the product being essentially commodities
- five serious worldwide suppliers are estimated, three being existing food equipment suppliers to laboratories and two being start up businesses.

It is likely that the worldwide market for the product will work out to be about $600 million annually at end user pricing, assuming that initial peaks will not coincide in each market. A serious participant in the market will need to have an annual sales aspiration of $100 million after five years, or $80 million after distributor margin is taken into account. This might correspond to 80 000 units at a price of $1000.

It is unlikely that prospective customers will start making this unit, as it will be a very small part of their purchases. No substitute technologies are on the horizon at this stage and current suppliers to the laboratory market, which are likely to be the main suppliers to the emerging market also, have high overhead structures and are considered unlikely to compete aggressively based on price. The marketplace, thus from this brief overview, looks as if it might be quite hospitable.

One issue not quite tied up is that of suppliers to the marketplace. The product can be made anywhere, but for the specialised sensor chip, and it might be in the company's interests to licence the chip to a competitor, one of the broadrange food laboratory equipment companies, to ensure volume demand. The action of potentially aiding a competitor may increase competitive advantage, and non-overlapping distribution channels may be available. The competitor might, for example, be planning to sell direct while the new company would use distributors, as having an international direct sales force would probably be out of the question.

On the issue of key success factors, obviously distribution is important. Being able to gain coverage through independent distributors in each major market is essential, as is having access to their service and support networks. Otherwise the direct capabilities of the major manufacturers could prove decisive. The regulatory climate in each country is also critical, but worries here might be eased by all major governments having made commitment to implement legislation within 30 months. This timing might be right in the context of needing to finalise product design and test, and sign up key distributors.

Technology change is always a risk. In the medium term the state of the art chip design will be sufficient, but the company will need to keep making products obsolete before another firm does. Apart from anything else, it looks as if this is an interesting market to which it will pay to continue to commit R&D resources.

This with a bottom up design approach.

Refining the concept of the product is ongoing; its specification includes that it will be self-calibrating, need no regular maintenance, be sealed and easily cleaned, but yet have the ability to sample the ambient environment. It also needs to be aesthetically pleasing. Should a range of colours be offered, or would the inevitable extra inventory requirements cause this to be more trouble than it might be worth?

For 80 000 units annually good tooling can be justified to give a unique styling to the product. Initially, different models for retail and for restaurant use might have been considered, but the buyer behaviour issues and the buyer needs are such that one model is appropriate. The product must also be designed for ease of use by the distributor, so straightforward servicing is essential. Good documentation for service technicians, visually detailed with many photographs to minimise errors, is particularly desirable. In supporting distributors, demonstration equipment could be provided and artwork supplied so that brochure material is easily translated. Matters must be made easy for business partners!

Selling a product internationally requires a design for the different electrical codes associated with each country. It might be slightly cheaper to use dedicated power supplies for each voltage region, but the simplicity of having a universal input supply requiring no switches may well more than justify the extra cost. If switches are incorporated, experience has shown that too many end users will set them up wrongly, with damage to the product and hassled distributors. The answer here can be for the distributor to perform the set up operation, but this again wastes time. The distributor should ship directly to the user site, with the store manager able to get the device operational. Knowing this has consequences for the clarity and language independence (where possible) of user documentation.

An area of product design easily overlooked is packaging design. The product is being designed for international shipment using a variety of carriers, and will be subjected to various drops and bumps as well as some extremes of temperature and humidity. Designing to protect against these conditions while still making the overall package attractive is the challenge.

Brochure material must be designed carefully. A product such as this is a distress purchase which the buyer ordinarily undertakes in order to meet a government imposed regulation rather than an enthusiastic purchase which can boost business. This attitude must be counteracted by putting the situation in a positive light, perhaps by pointing out that customer's health is being safeguarded by the new equipment, and by trying to build sales before the product become mandatory. Major retail groups must be the first in their country to adopt this new safety system, and thus put over the product in a favourable light where possible.

In the production area, there are two issues. One concerns production of the proprietary part, the sensor, and the other concerns production of the overall unit. Making the 'make or buy' decision for the sensor is quite easy, as investment in a dedicated semiconductor facility for small volumes of product is not justified. Given that this part is also being sold to a strategic partner, two sources might be selected and performance compared, giving two windows on the development of the all important process technology.

In the case of manufacturing of the instrument, a total outsource approach with one manufacturer, involving test and shipping of finished product direct to distributors, complete with documentation, might be chosen.

Managerial resources must also be considered. Working with distributors takes time, particularly in the early days, as they often need to be oversupported so that they have no difficulties with the product and are not distracted from it, being tempted to spend time on lines which are easier to sell than new products which require a special 'missionary' sales effort, even if they are required by law!

The overall approach has to be one of integrated design, taking in all aspects, and broadening the design of the traditional, narrowly defined product to take in many aspects of the broader business.

Having gone this far with the bottom up approach, the top down approach can be revisited for consistency checking. At this stage some contingency strategies can be considered.

If enough money cannot be raised to go into production, what is the best option? One approach is to proceed with getting the sensor chip manufactured, and work with a partner. Another is to sell out the whole venture at a relatively early stage, realising some gain if not the ultimate potential gain, but also saving the effort and risk needed to develop the full business along the lines envisaged. It might be decided that world coverage was not something to take on, even with distributors, due to management resource constraints. In this case the rights for Europe for the final product could be retained, and the product licenced outside Europe, with the sensor chip also sold separately.

The approaches outlined here cover some of the main strategic issues associated with business design, but not all. Every business must be equipped with information and control systems, have appropriate quality measures in place and have policies for ongoing technology development and for looking after its brands and other intangible assets. The organisation of the business must also be designed, considering both traditional approaches of task division among individuals and re-engineering ideologies focused on teams and processes. The internal issues are discussed in Chapter 4. Underlying all business design is the issue of finance and this is treated in Chapter 5, along with the specifics of raising money, discussed in Chapter 6.

Internal business design issues

Having mapped out the relationships and the heart of the design approach, key aspects of the entity being designed now need to be considered.

From the viewpoint of the entrepreneur, the important thing is to address these internal design issues at a very early stage. Getting the right attitudes and culture into an organisation, and establishing the correct organisational structure principles, avoids the costly issues of change management at a later stage.

There is a large body of current ideology associated with manufacturing and other areas within the organisation, and it is important that the company is imbued with best practices in these areas from an early stage. Installing information systems and setting up organisational structures is also best undertaken as early as possible. Putting full scale implementations of systems into a very small organisation may be impractical, but the important task is to ensure that no action is taken which will make future implementations of these systems unduly difficult as the business grows.

4.1 The information/control/quality systems

4.1.1 Information and control systems

Information is central to an organisation, a counterpart to the nervous system in a human. The information system allows us to know what is happening in a business, and usually also has a facility for processing and filtering the raw information to give reports of value. But it is not solely concerned with internal matters; a company needs to have information on external matters also. At an operating level, it may be coupled via computer systems using electronic data interchange (EDI) with suppliers, distributors and customers, working increasingly in just-in-time arrangements to limit inventory.

The objectives of the information system are really concerned with tracking what is going on in the business and in its environment. Information needed includes:

Orders and likely orders

Tracking orders is usually relatively easy. In terms of tracking likely orders, the buying patterns of customers and distributors need to be known with their existing stock levels and possible economic influences.

Marketing in general

Information is needed on competitors and potential competitors, and on suppliers. Customers must be monitored, with a watch kept on their health. Healthy customers can receive more attention, while those in difficulty may be credit risks.

Production

Details of what is being produced, utilisation of capacity, limitations and location of inventory are required. Within the procurement function the company needs to know where its materials are and be able to plan materials usage in advance so that deliveries from suppliers can be scheduled.

Service

Good customer feedback is necessary for monitoring reliability.

Product development

Knowing what is happening in product development is useful in forecasting production volumes for some time ahead. R&D scheduling is an activity which is rarely done with precision, but at least some estimates are required.

Resources

Human resource capabilities, training and other details need to be known along with information about physical assets.

Quality issues

A quantitative approach is needed to quality concerns, measuring the costs of any failings in these areas, product quality and issues such as service delays.

In this description a financial information system has deliberately not been listed *per se*. The financial aspect is ideally inherent in all information system modules.

The information system has both quantitative and qualitative aspects; there is often merit in trying to make matters as quantitative as possible. For example, customers can be scored on a points basis in terms of key issues such as growth rate, new product introduction and creditworthiness, rather than relying on mere words.

Information systems have become more important to businesses as their capabilities have increased. The motivation for the increased attention being paid to this area comes from several factors.

Extra complexity

Businesses now tend to be more complex in almost all areas. Customers want wider varieties of products in broader colour ranges and manufacturers and retailers carry large numbers of stock items. Mass customisation of product may be required. Giving this extra variety to customers requires flexibility, which increases the demand on the information system.

Marketing requirements

For travel companies the reservations system is a central competitive tool. Being able to give a reservation confirmation quickly is probably essential, and having good capabilities in this area offers power for forming favorable links with other travel providers. Increasingly a company's distributors and end customers need access to their system; they need to see stock levels if they are to judge their own best stock levels, and they need to know likely ship dates on custom orders. Many companies demand compatibility with EDI systems on the part of their suppliers.

Better utilisation of assets

Given the focus on performance of companies, a particular concern is with reducing inventory. Coupled with the continual need for additional variants in the manufacturing process is a need to have information to allow scheduling of orders and deliveries.

General competitive pressure

Competitive pressures are a natural driver in the information context. Response needs to be as quick and accurate as that of competitors, and in other operations aspects of the business the company is always ranked against competition.

A desire to prevent problems

Tied to the overall quality ethos which all companies must embrace is a need to measure continually many of the key operating parameters of the business. In this way one can see if an operation is gradually drifting out of control. If, for example, doors have to be 1.98 metres high (plus or minus 2 mm), then the measurement system can be set to alert staff when production begins to go outside a 1 mm tolerance. In this fashion the equipment can be checked and a tool changed in advance of producing a large amount of out of specification product. Statistical techniques such as SPC (statistical process control) can be used to determine the significance of results and to determine when to act.

A need to monitor supplier performance

Most businesses recognise that working with suppliers on a partnership basis is the only way to achieve long term success and avoid, in many cases, the need for investment either in manufacturing components or inventory. To manage the supplier base requires quantitative measurement of performance, ranging from conformance measures in terms of product attributes to delivery times and order responsiveness.

As well as the operations management aspects of the information system, management needs to be given an overall view of the company without being immersed in a morass of detail. Exception reporting monitors performance against standards and seeks to sound the alarm when the process begins to deviate from planned performance. Executive information systems (EIS) have recently become available from a number of vendors and allow managers to get summary reports, but also to analyse numbers in detail, from a workstation. The ability to get the overall picture; while also being able to access detail, is the attraction of these systems.

A company board receives information in a highly distilled form, as does top management. The attraction of these EIS systems is that detail can be obtained without the need for a special request for an investigation.

What is being presented here is really a control system (Figure 4.1) where management is ordinarily reactive to reports received. Engineers and mathematicians have spent much time in the modelling of control systems, and the basis of this type of system is similar in its fundamentals to a servo mechanism as might be used in a piece of industrial machinery. The theory of control systems which has been developed in this context can tell us that stability is a very desirable objective, and that this is usually ensured by having fast response times.

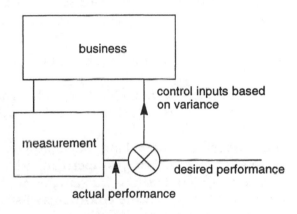

control system view of operational management

Figure 4.1   The control system analogy

In a business control system, forecasting sales and being able to react quickly to the forecast is probably the most important aspect.

Having an inventory pipeline through distributors to the end customer can be dangerous, as it may amplify any swings in demand by the customer. If end customer demand is slow, distributors may be overstocked; as demand picks up they may be fearful of overstocking and not order. Then, possibly simultaneously, the plant will be flooded with orders as the distributors attempt to supply customer demand and build their inventory. A moderate fluctuation in end customer demand can thus be amplified greatly (Figure 4.2).

Qualitative information needs to be obtained from end customers to determine how the products are being received. It is for this reason that manufacturers who typically sell through distributors may keep

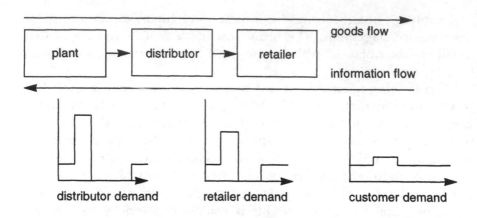

Figure 4.2 Small variations in customer demand may lead to large swings in plant output

one territory in which they sell directly, providing this direct customer liaison.

Designing an information system has thus both formal and informal aspects. Some of these aspects will lend themselves to automation, and others less so.

The formal aspect can be reliant on computer based systems, although these are rarely perfect. Just specifying what is required can be very expensive as every business has unique needs which cannot easily be accommodated within the constraints of standard packages. Expensive customisation may be required. In manufacturing, MRP (materials requirements planning) systems have gained popularity, but linking these systems with others in the company is often not easy.

Database techniques have advanced significantly in recent years, with most modern programs being structured around a database. Increasingly, one speaks of a repository of information which can be accessed selectively to generate the required reports. If a repository of all transactions and all production batches, as well as all shipments, is maintained, queries can be addressed to this store to generate information on accounting, delays in handling, customer shipments etc.

The business also needs to take a less quantitative approach to information as it is managed by people and codification of all the subtleties involved is impossible. There remains a need for individual judgment and for 'management by walking about'.

Information systems also have implications for organisation design. One consequence of better information systems has been that an individual manager can control more subordinate functions. Being able to review a number of reports on each centre is easier and faster for that manager than extracting the information from subordinates on an individual basis. Thus banks operate with large numbers of branches reporting directly to headquarters, and in general the number of layers of management can be reduced.

One model of the managerial role shows the manager as really an information processor, making decisions on the basis of information supplied, and able to get overloaded quite easily. This model may appear rather clinical and offering a limited interpretation of a manager's job, but in the design of organisations and in the design of information systems, it is relevant. Information systems must be designed to give the manager the relevant material in the most appropriate presentation.

Speed has become a major competitive weapon, and better information systems allow more effective management and faster response to customer enquiries, often enabling the sale to be concluded on the spot. A somewhat missionary approach has to be adopted to ensure that delays are eliminated. There is no reason in most organisations why full management accounts cannot be produced on demand, in real time if required.

There are points of debate concerning all accounts. Some contracts are ongoing and accruals and provisions are required. There is, however, no reason why these factors should make daily accounts irrelevant. Certainly there can be daily fluctuations, but comparison against comparable dates a year ago and some weekly aggregation can also help. The point here is that the response time of the organisation must be speeded up. If a problem occurs in January in a conventional organisation, it might appear in papers for a management meeting or board meeting in mid February. An anomaly might be dismissed as a fluke if it appears in only one report. If, however, it is not a fluke, management is not really aware of what is happening until mid March, by which time two months have elapsed. Many companies now get flash reports of some key variable such as sales or gross margin, although there is no reason why the complete accounts package should not be available. Business runs as a continuum, and must not be tied to a period based mentality just because tradition has favoured this approach.

Speed also needs to be deployed in serving customers. Airlines are usually very good at this, but think of all those free telephone numbers that say 'allow three weeks' for subsequent delivery of printed material. Do they really need to wait three weeks, or is this just the way they have always done it? If somebody calls a telephone number to get information on a new car or investment fund, is there any reason why the letter cannot go out that evening, to be received at most two days later? Here is a qualified prospect being treated poorly; not good business sense! Likewise when ordering a magazine, why cannot the first regular issue arrive as soon as it is published. Why are several weeks needed? Being oriented to speed of fulfilment throughout the organisation is important, the crucial question being 'why are we waiting?'. In an insurance company if a quotation is given on the spot the conversion rate is much greater than following conventional routes. Speed comes into manufacturing planning and also to R&D work, and these aspects are considered in subsequent sections.

In some studies, the business environment has been analysed to show likely levels of turbulence, with different approaches being described as suited to various industries. In practice all industries have now to be regarded as turbulent, or potentially so. Advances in information systems make possible new ordering patterns and allow new entrants from remote locations to serve markets by telemarketing. The businesses traditionally treated complacently and viewed as 'cash cows' are often subject to as much change as the emerging business areas known to be turbulent. An ongoing program is needed to imbue these units with a system of self examination and benchmarking to improve quality levels, service times and all the other measures, not least for defensive reasons.

There is usually a company way of doing things, and it helps if all subscribe to this. Procedures manuals are less popular than once was the case, but their message is useful in coordinating the response of all in the business to a certain situation. It helps if the organisation and people within it react predictably.

Tight control of a limited number of key points is always needed. Agreements which can bind the company cause particular problems. An example is where Westinghouse was committed to supply fuel for reactors at a fixed price, involving a huge open liability. Tight control must be exerted in these key areas, while allowing tactical freedom to local management.

4.1.2 Quality systems

Just as the financial system is really subsumed within the overall information system design, so also is the quality system. Quality is a value inculcated at all steps of the process, manufacturing or service, seeking to ensure quality at source rather than having quality 'policemen' at various points throughout the organisation. Many quality issues are really addressed at both the market analysis stage and the design stage. David Garvin (GARVIN, 1987) has given eight measures of quality; commenting on each of these.

Performance

Performance really has to be measured in relation to the customers' expectations. Demands will be less for a product known in advance to be cheap and cheerful than for an expensive, high performance product. A zinc-carbon battery is expected to die before an alkaline battery. The key is meeting expectations. The basis of market segmentation is often performance.

Features

Feature selection is again a market related issue. The market can be confused by excessive features; telephone systems are an example of this phenomenon.

Reliability

Sometimes a matter of life or death, reliability is, on other occasions, a matter of minor inconvenience. Mean time between failure is a useful concept for electronic products.

Conformance

Here the consideration is whether the product meets the specification, explicit or implicit. In industrial bids, a very quantitative specification is given where with each point a contractor claims conformance or seeks a derogation. If the product does not conform upon shipment, then there is a problem.

Durability

Durability could possibly be subsumed under conformance or performance. However, the specification is rarely written tightly in the case of most consumer goods and durability is a qualitative judgment relating to the lack of fragility of the product.

Serviceability

This is usually measured by the mean time to repair. If a travel reservations system goes down, a five minute drop can possibly be tolerated but anything longer could cause problems. The business is designed around customer requirements, taking in product design areas where diagnostic capability might be built in, service force and company policy on the amount of repair material in the field.

Aesthetics

What looks good often is good, and customers have received this conditioning. In many cases where the internal workings of a product are not obvious or not easily understood, the tangible appearance of the product becomes important. Software vendors are big spenders on good packaging for mass-market software, as this is usually the only product attribute visible to the customer.

Perceived quality

The important aspect here is that customer perception of quality may be different from the company view. Garvin (*op. cit.*) cites the example of the telephone company which assigned high importance to call set up time and minimal importance to transmission noise during the call. Customers had a different set of priorities.

Quality control in manufacturing is thus the proverbial tip of the iceberg. Quality is really an element of integrated business design, taking in marketing, design and service as well as manufacturing.

In marketing, customer requirements need to be understood. When selling a product whose cost has minimal impact on the overall cost of a customer project but whose performance is vital, it needs to be sold on quality with significant freedom in pricing. This is often the case for suppliers of key components to military programs. The supplier of specialised components for mainframe computers is in a similar position, where the sale will be primarily based on reliability.

Most quality issues are determined in the product design area. If a product is not designed correctly, quality is impossible to achieve during manufacturing. Particular attention has been paid to these issues in Japan, with the work of Taguchi emphasising defensive, robust design if quality is to be maintained under a wide range of operating conditions, and introducing the concept of signal to noise ratio, where the signal represents the desired performance and noise represents the perturbations which can occur in the various stages from manufacturing through to final customer usage (EARLY, 1994). Design must be done defensively, with diagnostic systems to show what is wrong or to give warnings. Many electronic systems can undertake self checking. In service applications, reality tests can be used to check whether a number being entered is likely to be correct. Systems must be designed that arc robust rather than finicky, and that do not require constant adjustment in the field. Designing the aesthetic aspects well will often prejudge the issue favourably.

In the manufacturing area, quality systems must be designed for incoming materials and parts, for work done in house and for product shipment. The starting point in these areas is the belief that in a perfect world all products received would be exactly to specification, and this specification would be adequate. Barriers to this state of perfection need to be systematically removed. This is a slow and painstaking process, but many companies are showing that it can be done. In the plant, processes must be designed defensively to minimise the risk of errors. Getting rid of random defects, such as poor solder joints in the electronics industry, is probably the most difficult task. A mixture of prevention, such as in training policies and in ensuring cleanliness of parts, and inspection and testing should be used.

It is important to catch defects at an early stage. Figure 4.3 shows the increase in costs that can occur as a defective item makes its way through to field use.

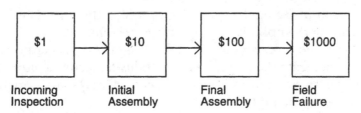

Figure 4.3   Extra cost of late failure diagnosis

FMEA (failure mode and effect analysis) techniques are useful when designing products and processes. Eliminating the obvious ways in which failures can occur, or minimising their effects through defensive techniques such as redundancy of systems, is essential in any defensive design. Knowledge needs to be updated continually, learning from failures which occur both internally and in comparable cases. The aircraft industry is usually considered to be a good benchmark for learning from failures and sharing knowledge.

The key aspect to be communicated about quality is that quality consciousness must permeate the organisation. Quality is not something to be enforced by a brigade of manufacturing police. Indeed by defensive design of product and of the manufacturing process more can be done for quality than by large numbers of inspectors. The importance of catching quality related problems early, i.e. at the market research stage, at the design stage or in the early stages of manufacturing is demonstrated by the typical costs of failures at various stages of the manufacturing cycle.

Quality is very much an issue of management, running right through the organisation, and ISO9000 and similar standards are mainly geared towards ensuring good management, with good record keeping, good audit trails and good reporting and responsibility definition. If quality is seen in quite a broad light, these same techniques can be used for managing the approach to environmental (e.g. BS 7750 standards) and other regulatory issues.

A key tenet here is rigorous attention to procedures. This ensures repeatability, and avoids the need for continued testing of products or services. There is a focus on documenting all steps to ensure traceability of product, and to ensure that any tests undertaken with equipment can be referred back to the standards.

The TQM (total quality management) approach (MUNRO-FAURE and MUNRO-FAURE, 1992) extends to dealing with the customer, in terms of verifying needs and requirements. MADU and KUEI, 1993, have also introduced strategic quality concepts.

Modern approaches to quality emphasise the associated human issues. The rigid emphasis on procedures is complemented by seeking broad employee involvement in establishing these procedures, and in seeking to delegate the quality responsibility to each individual. There is an emphasis on prevention of defects rather than remedy, with 'right first time' being the objective.

A key part of systems for information/control/quality is the use of periodic reviews or audits. An audit is not limited to the financial audit, which most companies undertake, and may be performed on marketing capabilities, or the production efficiency in various plants. What is needed is a situation of continuous improvement, benchmarking the way certain functions are implemented.

Quality does not come free; there are identifiable analysis and prevention costs, and measurement requires time and interpretation. The expectation is that these costs will be less than the failure costs, which can often be very large. Failure of a life-dependent component, for example in an aircraft, is simply not acceptable, thus the heavy emphasis on quality in industries serving this market.

4.1.3 Benchmarking

It is often valuable to be able to compare the business operations of one company with those of others. Parallels in the same industry can be used where information can be obtained on the operations of competitors, and sometimes comparisons can be made with or at least ideas drawn from, completely different industries. This process is usually termed benchmarking, (e.g. SHETTY, 1993, OHINATA, 1994) where processes are compared with the best in the industry, or with the best practice across all industries. Some industries have developed skills in specific areas in which they have been particularly challenged, and others can learn from these. Examples of industries which are recognised as setting standards include

- distribution logistics—food retailing; other industries learning: other distribution businesses
- maximising use of fixed assets, segmenting buyers by time sensitivity —airlines; other industries learning: hotels, automobile rental, telephone companies
- franchising—fast food; other industries learning: other local service businesses
- innovative financing approaches—transportation assets; other industries learning: telecommunications, power plants, other infrastructure funding
- management of large SKU (stock keeping unit) inventories— electronic parts wholesalers; other industries learning: healthcare product distributors.

In other cases it may be particular firms which set the standard for benchmarking. Some firms may excel in employee skills development, others in good operations management. The approach of benchmarking has become a key tool for various management consultancy firms (e.g. ARTHUR ANDERSEN, 1993).

4.2 The new operations ethos

Manufacturing was, for a time, the neglected area when it came to corporate strategy issues, at least in the western world. It has, however, increasingly become regarded as the key area for gaining competitive advantage. Manufacturing and its associated logistics operations usually account for the bulk of the costs associated with the finished product, and flexible manufacturing allows the supply of differentiated products, optimised for the needs of specific customers.

Manufacturing also accounts for the bulk of the fixed asset investment in the business, and is not easily changed, so investment in manufacturing effectively binds a company to a certain strategy for a number of years. Perhaps the key article in reawakening interest in the importance of manufacturing in corporate strategy was SKINNER, 1969. An interesting perspective is also given in HAYES and PISANO, 1994.

The 1990s have seen considerable emphasis on the rewriting of much of the conventional logic associated with manufacturing, or more generally, operations. The old approach was to have large inventories in place in order to ensure production continuity, and to use large batch sizes for maximising machine utilisation (minimising the time allocated to make ready and other set up procedures). Also, the work was to be divided into operations which were deskilled in so far as possible, with a hierarchical approach to supervision.

The new approaches stress:

Flexible or agile manufacturing

Special purpose equipment, optimised for one product, can help give cost leadership in that product, but in that product only. In a turbulent environment, it is very unlikely that this one product will not have to change in some respects over time, and production equipment may then have been optimised for the wrong goals. Paying a little more for flexibility in manufacturing can mean greater strategic flexibility, and the ability to manufacture a wider range of products, providing the opportunity for product differentiation if the broader strategy calls for this.

Flexibility can also be gained by using a mixture of full time and part time staff, with the part time resource being available to meet peak demands. Arrangements that readily permit overtime working in times of peak demand coupled with an ability for this overtime work to be traded for leave in slack periods are also of value.

Tapered integration, where the base load of work is done in the plant and subcontractors are able to absorb peak demands is an option. The difficulty here is that good subcontractors may find it difficult to handle a demand with accentuated fluctuations, and peak demand periods for all their customers may coincide. Some creativity in the mix of work undertaken should be deployed, with background tasks of lower priority accepted on tighter pricing, on the basis that they can be pre-empted by higher margin, more urgent work. Of course, steps can be taken to minimise peaks through better communications with customers, seeking to improve forecasts and to eliminate the end of period syndrome which affects many industries, as customers seek to close off their reporting periods—monthly, quarterly or annually—with a burst of activity requiring a corresponding burst on the part of their suppliers.

Minimal inventory holdings

Suppliers should deliver within very short time windows (just in time), and likewise manufacturing should be scheduled for delivery to the customer with a precise time indication. Avoidance of inventory minimises the full costs of inventory handling, which include obsolescence and deterioration risks, costs of warehousing and multiple handling and the extra packaging which may be required if inventory is to be retained for long periods. When minimal inventory is held items can be stored adjacent to machines, clearly visible to operatives of the job being undertaken.

Reducing batch sizes

Optimising machine loading can have a high price in terms of inventory costs and reduced responsiveness to customer needs. *The Goal* (GOLDRATT and COX, 1984) outlines the issues—traditional accounting measures focus on machine utilisation figures but customer needs are met best by small batch sizes with consequent loss of machine utilisation.

Empowerment, teams and responsibility for checking

The logic here is that operatives can be relied on to come up with process improvements, and that job performance can be optimised if peer pressure becomes the local control mechanism. This peer pressure means that the individual and the group can take responsibility for many quality issues, and compensation can be allocated, at least in part, based on group productivity.

Teams may be best placed to operate in cells where workflow is clear and inventory is conveniently located next to the manufacturing facilities within the cell. All aspects of the manufacturing operation for which the group has responsibility are visible, and thus problems are spotted more readily than in a large system.

Kanban systems operate to provide automatic reordering, limiting the amount of inventory required.

Continuous improvement

Team based approaches, with local control of production issues, are conducive to adoption of a series of minor improvements in the production process, which together can add up over time to very material savings. The emphasis is on taking decisions at the local level where possible, empowering operatives to make decisions, and tying rewards to team performance.

Mass customisation

Effectively making custom products without sacrificing the production economies associated with mass production, with the consequent demands on systems and controls. (See, for example, PINE, VICTOR and BOYNTON, 1993.)

The basis of these new approaches, taking in elements of lean production, world class manufacturing and other models, is putting customer requirements first, optimising the right parameters rather than the wrong parameters such as machine utilisation.

Introducing lean production also has effects throughout the industry. Suppliers are told to deliver according to a fixed schedule or, better still, given access to production planning details so that they know the required shipping amounts and times. Customers can also be offered better delivery estimates and greater responsiveness, thus gaining a competitive advantage if the company is the first mover in the industry and if customers are able to appreciate the advantages associated with this move. The lean approach can be extended profitably to permeate the entire business (WOMACK and JONES, 1994).

4.3 Technology

Technology may be viewed broadly as the core competence of the business., reflecting the knowledge base of the firm contained in its formal intellectual property assets such as patents and copyright, the applied skills of its people and the processes and techniques that it has developed over time.

In any business area, competitive advantage is the key and relative technology positions are important. The competition may be on the basis of product technology, where one company has a better product than competitors, or on the basis of process technology, where a product can be supplied more quickly or more cheaply or with more variants. The division between process technology and product technology can become blurred. Most products are now designed for manufacture, where improvements in process technology may be made at the same time at product technology improvements. Products can be self testing and self calibrating, desirable attributes from the viewpoint of the end customer and from the viewpoint of manufacturing and service aspects.

Technology is thus a resource, an intangible one, of the company. In designing a business one needs to plan for sustained competitive advantage, and technology is central to achieving this. Technology is also something that is developed over time. Research and development programs take time, and developing a new process technology is usually somewhat more complicated than buying a new piece of plant. Technology management is also not just for high technology companies. Most manufacturers should be continually updating process technology, and service businesses often derive competitive advantage from their use of process technology in the form of information systems.

A company's technology approach will, of course, be largely directed by the marketplace. It may seek to be the leader in product technology, or may emphasise process technology or service as elements of its approach to the customer. Maintaining a good position technologically requires looking after a range of issues.

4.3.1 Maintenance of product technology

Keeping up product technology is important, even if the company elects not to be quite the leader. Technological followers cannot survive solely by service and by process technology strengths if they do not have the ability to come out with the latest desired features within a short time of the leaders. This implies a continued need to upgrade product technology, by some combination of technology purchase and in house development. An in house capability is usually needed, even if it is limited to supervising contract R&D and looking after other technology procurement, and some measure of continuity in the technology function is essential. Having in house control helps. If the company is buying a technology license, then it will need in house capability, as technology transfer in a license deal involves the need for a competent recipient. One of the attractions of licensing technology, exploited very well by many companies in developing countries, is that the license can be obtained in the early stages of development. The company can then develop their own competences when the broader industry knowledge has been developed.

Managing the R&D capability is usually a different challenge to managing another department, with many R&D teams working very effectively as small units. In the classical piece *The mythical man-month* (BROOKS, 1974), the author points out that a task sometimes cannot be partitioned into man sized pieces, and that adding new people to a late software project can make it run later, with the extra people having to be trained and with extra communications required within the team. The chief programmer team approach (BAKER, 1972) has been cited as maybe the best way to run a software project, where the idea is that the leader behaves as a true team leader, working as well as managing, but with significant administrative support removing the more mundane managerial tasks. This chief programmer team seeks to supply acolytes to the master designer, and this model may be useful in other R&D areas. The emphasis in an R&D department is usually on speed and on low cost, and both these objectives are met by the use of effective small teams.

The DEC project for the development of the Alpha RISC (reduced instruction set complexity) processor is cited as an example, being essentially led by a small (ten person) team. This is because, or in spite, of its evident importance for the future of the company.

The ultimate in such teams is probably the 'skunkworks'. This approach holds that product designers work best in an isolated setting, if anything feeling slightly pressured by a competing approach. If a company can afford it, it can have a skunkworks running in parallel with a more regular design group (e.g. in *Soul of a new Machine* (KIDDER, 1982). Despite the obvious waste in the R&D costs, the company could gain in the overall analysis if the product gets to market more quickly. Many product design areas involve a race; a new chip or a new operating system appears, and an offering based on this new chip or system has to be available at least as rapidly as one from competitors.

Keeping research and development team members within the company is an issue. Skills at the forefront of most technologies tend, by their nature, to be very specific, and replacing a key member of the team, particularly at a critical stage in the project, can be very difficult. Motivation issues for R&D team members are probably more acute than usual, with a high need for achievement and a frustration with the company which does not supply the best tools for creativity. Hence the healthy market for engineering workstations and other design support equipment!

Location of the R&D function is also important, absorbing information from suppliers, customers and competitors, but without losing data to the competitors. While development teams can function well in remote areas, there is much to be said for being close to acknowledged centres of excellence. Some companies show a particular liking for small towns with good educational facilities to hand.

The relationship between marketing and R&D groups also needs to be designed. These departments often have a different group culture, with the R&D group composed of goal oriented people, who may sometimes have the goalposts set too narrowly. The objective of the R&D role is to come up with winning products, with winning being defined in the context of the overall health of the firm. While technical perfectionism may have to be reined in if the design is ever to be frozen for manufacturing, the designers' need for achievement is satisfied much more by products which are successful in the marketplace than by the curiosity product that broke new ground in some area but which is a commercial flop.

One conventional approach has been that the R&D function is the servant of marketing. The marketing people identify what the customers want, which is then addressed by the product designers. This is, however, an incomplete picture; a company needs to be designing what its customer will want some months or some years ahead. Technologists have a vantage point on the capabilities that will be available, and it is often necessary to take a bottom up, technology led approach in terms of seeing what is possible. Advances in manufacturing technology or in the service areas can also provide the lead.

Product design capability is often given an unduly narrow specification. For a new amplifier, power, gain and frequency range might be specified, as well as some other technical parameters. What should also be communicated are desired cost, the quantities expected, so that initial approaches to best manufacturing can be considered, the distribution approaches, which impinge on service requirements and whether self-test capabilities should be built in. All these other aspects are relevant to the product design function, and indeed the design of the extended product, taking in the packaging, advertising, support and all the other product attributes, should be in the mind of R&D people.

Many industries experience race conditions as companies jockey for position in a rapidly growing area. Technology will always be a component in this race, with the R&D effort required to develop a product several years into the growth of the industry being at least an order of magnitude greater than was the case as the industry started. Compare development tasks for a PC software product of the 1990s with the initial spreadsheet and word processor offerings of the early 1980s.

Speeding up design activities is a key challenge (TOWNER, 1994, WALKER, 1993). The idea is to have as much concurrent design as possible, and in some cases two different groups trying alternative solutions to the same problem. Figure 4.4 shows some of the approaches involved; the sequential approach has been described using sporting metaphors as the relay approach, with parallel efforts termed the rugby approach.

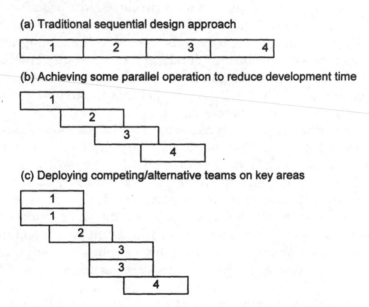

Figure 4.4　Approaches to speeding up product development activity

Protecting proprietary technology is important. Patents are the most obvious route, but they may provide ideas to competitors and provide the stimulus to allow them to design around patent claims. Putting appropriate language in contracts of employment and appropriate legends on documentation are also basic steps to be taken to protect intellectual property. A full treatment of these issues is given in IRISH, 1994.

4.3.2 Process technology

Viewed simplistically, most process technology can be bought. In practice, however, in house process technology can be very important, for example, a key edge in the confectionery industry is obtained from the use of suitable packaging equipment. Manufacturing processes must be optimised. As a simple example, it is usually preferable to design in snap fixtures rather than screws, assuming that both satisfy the strict product performance specification requirement.

Process technology revolves heavily around the know-how that is usually built up over time. Experience is codified. Know how is essentially a trade secret and it is rarely patentable. If it were, disclosure would not be appropriate.

Process technology is also tied very closely with staff training, and getting production teams to understand the details of the process and its inevitable quirks.

4.3.3 Overall technology policy issues, standards

A point of debate arises concerning standards. In many emerging industries there are standards battles with a clear cut victory for one party. Examples include eight track versus compact cassette, BetaMax versus VHS, MCA versus EISA, MS DOS versus CP/M and Unix versus proprietary operating systems. It is important to be on the winning side in battles of this sort.

Intel licensed widely its 8088/8086/80286 processor family resulting in wide acceptance as the processor of choice for DOS operating systems. When this was established, the 80386 and 80486 (improved versions with similar basic architecture) reverted to being again a single source product until court actions from early 1992 allowed competition. From 1993 until 1995 the Pentium was effectively, again, single sourced. Thus Intel had a single source dominant position in a standard where the market development was undertaken in association with others.

Technology out-licensing is an important tool in the game kit. Used carefully it can help build a standard, around which is created an area which the company is well placed to lead. Used with less care, it can bring other entrants into the business which, having been introduced to the various commercial aspects, can then develop their own technology and emerge as competitors. Many Japanese companies did very well from licenses obtained from US companies after World War II.

For the licensee, technology licensing is cheaper than developing a product, particularly when operating in a different geographical area or if the licensed technology is not the main area in which competitive advantage is secured. The key time is during subsequent rounds. If the licensee is in danger of being cut off from the next round of technology development and if effective ownership of the standard reverts to the leader, the advancement path is cut off. This will be permissible in some industries and not in others. Most industrial vendors need to have multiple sources, which provide an umbrella of sorts for one or two followers in the particular product line. If the standards issue is less real, then the licensee has learned a lot and may be able to develop its own products for markets it has already begun to serve using licensed technology.

The time to market issue is particularly critical with short life cycle products. With life cycles in many areas down to less than two years, a three month launch delay can allow others to set industry standards. The temptation here is to throw money at development activities to make them go faster, which is often counterproductive. A corollary to Brook's law is that 'adding more people to a late software project makes it later' and this probably applies across a broader range of development tasks than just software design!

4.3.4 Being first

In the commercial world, being first to market with a product is no guarantee of success. Having the resources to ensure that the product achieves wide industry acceptance is often the greater challenge. Being first also makes life so much easier for rivals. They now have a clear target and renewed confidence for their development efforts, and many of the regulatory hurdles and development issues for complementary products needed for the market to develop may have been overcome.

In the rush to be first the leading company may also have omitted a large number of possibly minor product features, so that the initial launch is eclipsed very quickly by a new generation of more reliable products with these added features.

Some of the issues associated with converting a technology lead to commercial success are treated in ZAHRA and BICKFORD, 1994 and FOSTER, 1986.

4.4 Brand management

As central as technology policy is the whole issue of reputation, which is tied in with the brands operated by a company.

Maintaining branding policy is particularly important. The central issue associated with brands is trust; customers buying a branded product expect to get quality and an element of reassurance and perceived consistency and reliability. Building respected brands requires significant time and effort, which of course translates into money. Brands can be valued at a nominal sum, or it may be appropriate to try to value them at something close to market value. This is of an art-form, as calculating the exact value will involve elements of subjective judgment. Parallels may be available in other companies, where much of the premium value paid over net assets in acquisitions has been assigned to the value of brands.

Brands have to be nurtured carefully, and used well. Putting a company's brand on inferior products destroys much of the reputation associated with that product, and if there is a major failing, the reputation of the brand can become irretrievably damaged.

In all design approaches value should be added to brands where possible. Brand extensions can sometimes be used very successfully, but caution is indicated. Likewise when licensing out a brand, consistency must be ensured. Some brands, well established in certain parts of the world, have been debased by careless licensing in others. With increased international travel, what may have been acceptable in earlier times can now be a major disadvantage.

4.5 Recruitment, team building, corporate culture and organisation structures

4.5.1 Recruitment

Building a management team is the key challenge for the entrepreneur seeking to develop a business. The company for many purposes is its people, and value will be assigned to it based on the achievements of its managers and other employees.

Getting good people to join small firms is not easy. Larger organisations can offer initially greater prestige; fellow guests at the cocktail party will have heard of them, whereas the reaction to an unknown small company on a remote trading estate or in some incubator unit is much more questioning.

Getting good people into small companies requires good selling and charisma from the entrepreneur, and the prospective recruit must see the advantages, namely an opportunity to

- do very much better financially, particularly if some form of real equity participation opportunity is available

- learn far more across a broader range of functions, given the small team nature of the work
- be very readily identified with success in creating a business.

These are some positive aspects. A start-up business needs people who will probably gravitate to that sort of environment, people who will be comfortable with long hours for deferred and risky, but potentially large, reward and who will be flexible about tasks to be undertaken.

The approach of the entrepreneur, in building a management team, must be to over recruit if this is possible. The ideal candidates are those with the potential to grow into managers within a larger company. It is much easier, if slightly more expensive in the short term, to recruit on this basis, as development can then be as much as possible from within. It will, of course, be necessary to bring in people at senior level as the business grows, particularly if one needs to sell internationally, but training a team and letting it develop will make sense.

The entrepreneur must also be honest and ask is he the right person to lead the business forward. The answer is very often 'yes', but it is appropriate that the question be put. If the objective of the entrepreneur is to maximise the value of his shareholding, then he needs to hire the managers best capable of maximising that valuation. A logical consequence of this analysis is that he may realise he could be better deployed as chairman and head of research and development, rather than as chief executive. It may be too risky to recruit in an expensive chief executive at an early stage, but it is something to which the entrepreneur should be open. If the objective is to run a one man band and enjoy the power and status associated with this, then the question probably does not arise.

The wise entrepreneur will also be choosing a potential successor, in the event that he wishes to retire from the business or to be able to devote time to building up a larger group in due course. In either event, gaining independence from day to day management issues is essential, and planning for a structure where delegation can take place is a key element of this. If one is creating value within the business, then it is important that the business is not unduly dependent on any one individual.

For an entrepreneur seeking to put together a team to secure venture funding for a new or developing business, having the major bases covered is important. These are usually the selling and product development activities. If production is a key issue, then covering this area is also necessary. Having financial skills within the team is desirable, but accountants do have the advantage of being much less specific when it comes to the job which they perform, and recruiting a good accountant is usually a somewhat easier process than recruiting specialists in rather arcane areas of technology.

Over recruiting for the financial role is also a good idea. Once again, if the company is to be venture capital backed, a requirement may well be that a suitable finance director is recruited. Having obtained such backing, the position on offer will be attractive and the business should be able to get a very good candidate. From the candidate's viewpoint, the attraction is that venture capitalists have done much of the investigative work, and on the back of this he is able to make his judgment. Venture capitalists may well be pushing the company to a public listing, where the finance director is likely to play a key role and be rewarded accordingly, as well as having an enhanced CV.

In recruiting people for an emerging venture, the company is offering not so much job security, as training and experience. This is particularly suitable for individuals who take more of the responsibility for their own career path, and who seek opportunities which are more likely to lead to future employability.

4.5.2 Team building

Team building comes down to interpersonal skills which some entrepreneurs have in abundance and others lack, although they may command a grudging respect from their management team colleagues. Some will be authoritarian at all times (most need to be authoritarian some of the time if things are to get done), and others will favour a *primus inter pares* style where possible. There is no observable correlation between style and effectiveness of the style. It all probably comes down to personalities!

There is no doubt that team building can be assisted by the difficult environment under which many early stage firms work. Stories can begin to accumulate in considerable numbers, telling of heroic feats of travel undertaken by executives on sales calls or when installing equipment in a remote location. These are the tales that bond the team and which keep the corporate culture strong.

4.5.3 Corporate culture

Corporate culture represents a body of values held by the company, or more specifically, by the individuals within it. Often there is a company way of doing things. The emphasis may be on customer service, such as ensuring that the telephone is answered in two rings at most, or based around pride in technical achievement and in product excellence.

One challenge facing a company is bringing about a change in the corporate culture. *When giants learn to dance* (KANTER, 1989) is a book outlining the challenges faced by many large firms as they seek to become agile and reduce development times, as well as becoming more open to new ideas. Telephone companies in the US have a long tradition of universal service and experience of operating in a regulated environment, and it was a considerable transition for them to go into competitive business. A similar change was experienced in telecommunications businesses in Europe, initially with British Telecom becoming a commercially oriented entity rather than state monopoly during the 1980s and followed by telecommunications companies in most other European countries making this transition in the 1990s.

Some of the aspects of corporate culture in a bank are highlighted in JOHNSON and SCHOLES, 1993, where the organisational paradigm was a provider of stable long term employment, with an indestructable organisation. 'Avoid risk' and 'mistake = death' were standard tenets, and a high emphasis was placed on professional status and integrity. An early stage firm wants to emphasise its integrity, but probably wants to be seen as quite the opposite in other areas. It wants to be seen as an attractive place to work, but in terms of adding value to people rather than assuming the responsibility for their lifelong employment. And it wants to allow for measured risk taking, with penalties only for sloppy execution rather than for having taken the risk, portraying itself as flexible.

A strong corporate culture is something of a two edged sword. Moulded for the positive, it can embody an emphasis on customer service, on technical excellence or on having sales staff meet quota 99 % of the time. Negative aspects can manifest themselves in arrogance or a 'not invented here' syndrome, where the organisation is unreceptive to new ideas, or can be associated with an organisation going into shock when the groundrules are changed.

Within a company, serving different types of customers can cause culture shock issues. The manufacturer of products for military and aerospace systems may have great difficulty in providing efficient design and manufacture of products at low cost for the commercial marketplace. For a product company, used to selling directly to end users, it can be quite wrenching becoming a service company supplying mainly OEMs (original equipment manufacturers).

For the start-up business, there is a unique opportunity to shape corporate culture from the beginning. It is possible to get it right without having to endure later the difficulty of getting people to change attitudes. Mission statements and other statements of what the company stands for may evoke a cynical response from some, but they are a clear statement of goals. When a company is recruiting, it needs to recruit people who are compatible with the corporate culture that it is trying to develop.

Cultural issues are also treated in DEAL and KENNEDY, 1982, HUMBLE, JACKSON and THOMSON, 1994 and KONO, 1994.

4.6 Hierarchies and team approaches

Organisations are managed by people. Designing a business includes designing the organisational structure. In doing this responsibilities are assigned to people and structures set up to permit people to get information to make the decisions needed to fulfil their responsibilities. The small, early stage organisation may often function very well with a blurred understanding of the organisation structure, but can become overwhelmed when the business grows beyond a certain point. In particular, if it is necessary to bring in new management, tolerance of apparent chaos, no matter how effective in practice, may be quite limited.

In designing a structure one cannot adopt an overly mechanistic approach. It is important to recognise the key roles of leadership and charisma. Business is essentially a human organisation, and structural design must reflect this. It is often more appropriate to design a structure around the individuals in the company rather than constraining the people to fit a predetermined structure.

4.7 Traditional approaches and re-engineering approaches

Many organisational structures have been based on the principle of division of labour. The basic approach was to deskill the job as much as possible and get one individual to specialise in one facet of the task in hand, thus leading to high productivity. This approach was at the heart of productivity advances made since the industrial revolution, and the hierarchy of control needed to make it work effectively continues in most organisation.

Since the start of the 1990s the hierarchical approach has been challenged by re-engineering (HAMMER and CHAMPY, 1993) which recognises that excessive fragmentation means that the key issue in an organisation—the actual process—often gets overlooked, resulting in a lack of clear responsibility for the visible results of many sections of the organisation. The process lacks owners and, while people may be doing their invididual job well, the fragmentation resulting from handover between people means that the overall process suffers.

4.8 Traditional structural design

The basic thrusts of structural design have traditionally been the functional organisation and the divisional organisation, with various hybrids.

In a functional organisation, the chief executive has reporting to him managers who are responsible for functions and for resources. These might typically include some of the following: sales, marketing, finance, R&D, production, procurement, human resources and service/support. Each manager handles the activities within his department, having a number of managers reporting to him, and so on down the pyramid. This structure can be very straightforward. There is a clear line of responsibility to the chief executive, and reports can flow up the organisation and commands flow down the pyramid (Figure 4.5).

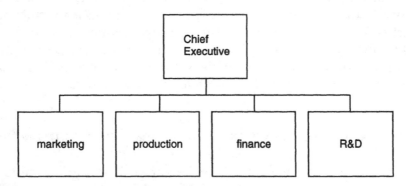

Figure 4.5   Traditional functional organisation

This organisational structure can be very satisfactory in many situations, particularly where the organisation is in a stable industry.

As an alternative to this there is the product group or market group structure (Figure 4.6). Here the chief executive has reporting to him the heads of a number of groups which are focused on key customer groups served by the business. Reporting also to the chief executive may be the heads of departments such as human resources and finance.

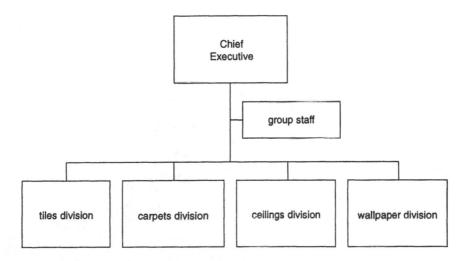

Figure 4.6   Divisional organisation by products (this could also be by geography, or by customer type, e.g. retail, wholesale or OEM)

Also possible is a geographical reporting structure, where the business is split by geographical region. This structure may be well suited to businesses which carry out similar activities across a number of branches or outlets.

In practice, none of these structures is likely to exist in its strictest form. Within the functional organisation will be found geographical and product-based allocation of responsibility, within the market group some central services, where typically production resources are shared amongst a number of product groups, or possibly human resources shared for consistency of policy across a group. Likewise, in an organisation with a primarily geographically-oriented reporting structure, some central services will usually be shared. In a bank, for example, the reporting structure for the branches may be organised geographically, but a central marketing, finance and human resources role may be maintained.

In practice, it becomes increasingly difficult to have a singular reporting structure. Usually something of a matrix structure is developed where staff answer to different superiors on different topics. Matrix management is confusing, but at least it formally recognises some of the dotted line relationships and liaison staff arrangements that develop in other settings (Figure 4.7). In all cases the objectives of managers need to be aligned with those of the overall organisation.

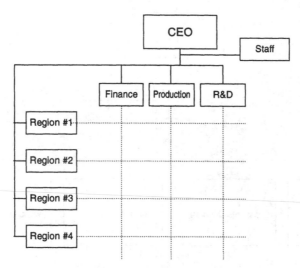

Figure 4.7   The matrix organisation

Recent work in the context of business process re-engineering has led to a focus on the design of processes as the starting point for design of organisational structure. It is logical to look first at the tasks that need to be undertaken by an organisation as the primary driver of any organisational matrix. The processes of an organisation need to be designed first; the organisational design, or at least significant parts of it, will then follow. Alternative organisation charts of various types are now appearing, as described in BYRNE, 1993.

Consideration of earlier structures has been based primarily on the classical view of managers, with authority devolved down and reports conveyed up. This approach is relevant, and indeed with better information systems its efficacy has often improved. The span of control, i.e. the number of subordinates reporting to a manager, has typically increased and this increase is credited to the superior reporting and data presentation capabilities of information systems.

4.9 The chief programmer team and analogous approaches

There is another aspect to information, and this is related to the fact that much of the information needed by people working on a particular task is unstructured, with people needing to meet informally to share opinions. This leads to the clustering together of those who need to work on a task in a particular cell or team. The chief programmer team, Figure 4.8, is an example of this approach, developed initially in the context of software design, where the most efficient approach was to have one talented individual as the nucleus of the team, with support staff grouped around. One person looks after liaison with the outside world, there is a secretary, an archivist, a team administrator and there are two general support staff. One can envisage this model being of value in other situations where the 'rainmaker' in a corporate finance business or the key sales executive dealing with major accounts need support.

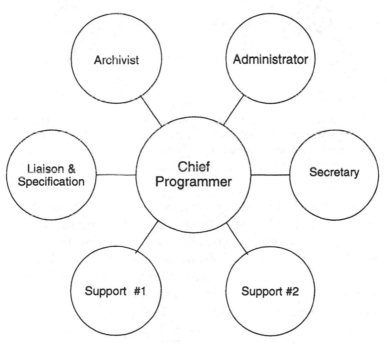

Figure 4.8   The chief programmer team structure

An information based view of organisation structure would lead to having a number of cells, with some overlap required for shared information. This structure can coexist with an overlay structure, which provides the authority and control aspects of an organisation, and which can provide some of the central services required by almost all companies. The challenge with organisation design with multiple relationships is to ensure that people are not in meetings all day! If the cells or teams can have clear objectives and purposes, this avoids much of the perpetual meeting syndrome. In many cases, teams are assembled and broken up in accordance with the demands of various projects. Figure 4.9 shows a possible information based structure.

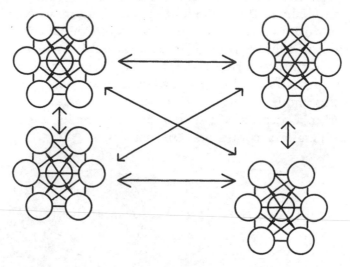

Figure 4.9   Information linkage between groups. A sensible design approach is often to minimise the information transfer required between groups, keeping teams compact and allowing a flatter organisation

The problem faced by most organisations, particularly successful ones, is bureaucracy and a resistance to change, reflecting the view that the organisational structure is sacrosanct and removing the attention of people within the organisation from the end customer. This difficulty can be particularly acute in functional organisations where the new product has to compete for attention with long established products, and necessary support is difficult to obtain. The larger organisation has a particular need for consistency. It may become policy that all managers of similar level have similar office decor, and a corresponding level of secretarial support. Customers require consistent service and employees need to act predictably in tasks where originality is inappropriate.

If consistency with the rest of the corporation is required, some clearly dysfunctional decisions may be needed which will give the wrong signals. Organisations also grow their corporate culture with a certain written or unwritten *modus operandi* in most situations. If one is to escape from the corporate mindset, then sometimes setting up a new division is required.

For IBM to enter the PC business, many of the rules within its established *modus operandi* had to be broken. This is a difficult task in any organisation, but particularly for IBM when the established approach had been so successful. IBM was a licensee for a non-proprietary operating system from a small company in Washington State, and was building its hardware with higher reliance on bought in semiconductor components and disk drives rather than using proprietary hardware. Sales channels needed to include dealer outlets as the original PC was also designed for use in the home environment. All these were major adjustments for the IBM mindset to take individually, let alone all together in one product. There was also a requirement for speed, with the whole development process taking eleven months. To achieve these tasks, IBM set up a separate division in Florida.

There are also situations where groups can be planned to be physically close together. In designing a plant, one might arrange to have product designers and marketing people with adjacent offices, to ease communications and facilitate working as teams on projects. Where close cooperation between the product design specialists and the manufacturing area is required, for example where a new product with simultaneous advances in product and process technology is being introduced, then having both groups located close together helps. In many situations task forces are set up to tackle a specific issue, and having close identification with the problem obviously promotes success.

Some groups can benefit from being removed from others. There is a school of belief that removes the headquarters from most of the trading business locations. The motivation for this is that the head office can be more objective in evaluating existing businesses. This, however, requires a strong divisional structure to be in place, as having to refer most decisions over a long distance is unlikely to add to efficiency. If manufacturing is straightforward and subcontracting could be considered at some time in the future, it makes sense to treat the manufacturing unit as distant and to formalise all communications with it. This means that a manufacturing documentation package is likely to be more complete and accurate than if there is the continued presence of a designer on the production floor!

Many businesses operate internationally, being forced to do so as competitors adopt similar strategies. The structure of the inter-national organisation probably depends heavily on the mindset involved. It may be truly international, a federation of national fiefdoms or it may be an exporter of product from its home country.

The true international organisation regards the world as its territory and will design, manufacture and sell in appropriate locations. The national fiefdom approach loses some validity with the opening up of trading blocs, but can still be valid. Here the national organisation can take product developed centrally and operate in each country much as a national supplier. The exporter will to some extent regard all countries other than the home base as peripheral, and is probably least well placed.

Improved communications facilities such as electronic mail and conferencing can make the first of these possibilities more viable than before.

Design of organisations also involves an objectives translation exercise. Every individual in the business has to be working towards some objectives and the net effect of these must be in broad conformity with the overall objectives of the firm. The objectives will typically be simpler as one goes down the organisation, and will ordinarily be reinforced by bonus payments or by the prospect of advancement within the firm, with corresponding increase in stature and remuneration. As objectives are translated, they are better expressed in positive terms rather than as constraints. If the corporate culture is communicated strongly, this can serve to set the context within which the objectives will be achieved.

There will, of course, be errors inherent in the objective translation process. The manufacturing manager is required to meet cost objectives, but also to push through a rush order for the special customer, although this will upset the cost measures. This is very much a problem in the traditional cost accounting approach, but is less so with the newer market-oriented cost accounting. Even then the value of a rush order for a key customer is unlikely to be recovered in the extra margin on that order alone, and some adjustment will be required if true justice is to be done.

Organisational design also has long term aspects. Most people like to be able to spend some time with an organisation, and it may only be possible to attract the best people if there is a prospect of real career development and long term employment. They may not take advantage of this prospect, but few people are comfortable working for the obvious fly by night firm or for the business with an excessive hire and fire reputation. Skilled and trained people are a key resource, and also one of the most inflexible in that they are hard to recruit, they are expensive and their services are required on a long-term basis.

The organisation must be designed to have strength in depth, with good managers in most of the key roles. As a small company it may hire people who initially seem overqualified for some of the roles, on the basis that their extra talents will come in useful as the organisation grows. Some companies have a policy of mainly promoting from within, while others will also recruit from outside. It is often attractive to be able to import new talent, if the new recruit can fit in and really understands the business of their company.

Interfaces between elements within the organisation are important. These are ideally modelled as customer/supplier relationships, where there is mutual respect and objectivity. In some cases outside competitors are introduced to evaluate internal resources and avoid the effects of an internal monopoly. The design department can be compared with external consultants, data processing functions with outsourcing contractors, production functions with outside suppliers and accounting functions can be compared with the outsourcing activities that professional firms are beginning to undertake in this area.

Perhaps the largest scale internal monopolies have been in government sectors. Irrespective of any ideological debate concerning state ownership of business, the approach of opening monopolies to competitive forces has brought about a new attention to benchmarking and to commercial success criteria.

Section III
Finance

Chapter 5
Financial concepts for the technical professional

Finance is perhaps *the* key issue in management of the business. Sometimes it is very easy to fall into the trap of thinking that if there is a good product, well sold, then the financial aspects of business will fall into place. Nothing could be further from the truth.

The fundamental purpose of a business is to create value for its shareholders. A business may well be assigned additional objectives such as job creation, economic development or other social goals, but it must create value if it is to be able to undertake these worthy additional goals in any continuing fashion.

Business is thus about value creation for the investors; they put in cash, and expect a return on this investment. Timing expectations for the return may vary significantly; some will expect to get a stream of income perhaps every half year by way of regular dividends while others will hope to get a considerable lump sum from sale of the business (or a significant part of it) at some time in the future. Naturally, a business often has to address issues going beyond the narrow shareholder interests. Staff, customers, suppliers, lenders and the local community will all have a vested interest in its success, but their needs are usually also best served if the company is successful and profitable for shareholders.

5.1 Financial statements

In this section the main tools of financial management will be presented, seeking to bring about a feel for the subject and how to spot some of the key issues from perusal of financial ratios. The objective is *not* to allow the reader to dispense with the services of a professional accountant, as the treatment here will necessarily be very brief and will not go into many of the issues associated, for example, with taxation.

The objective of financial management is primarily to monitor the movements of value within a business. The three important financial statements are the profit and loss account (alternatively known as the income statement), the balance sheet and the cash flow statement (or sources and uses of funds statement).

These statements will be presented in the context of helping to run a business. The focus is not on detailed bookkeeping, which is increasingly being addressed very effectively for most businesses by easily used accounts software packages, but on seeking to communicate the financial concepts in an engineering context. The level of numeracy necessary to understand financial statements is usually much less demanding than that required for the treatment of dynamic engineering systems, but there are conventions of language and concepts with which the entrepreneur must be familiar.

The income statement primarily shows what value has been created for the shareholders in a given period. Very simply, it represents the statement that

value created = sales (of products or services) – cost (incurred in order to sell these goods and services)

These items can be analysed in more detail. A manufacturing operation is used to illustrate this. A software operation or a consultancy operation would be somewhat simpler, but the manufacturing operation gives a more complete picture.

The sales line is usually quite simple. If the company sells 1000 amplifiers in a month at £100 each, then the sales figure (the terms turnover, sales and revenues are used synonymously) is £100 000.

In practice, however, the cost line is analysed in a little more detail. Typically, amongst costs there are

Direct cost of materials used

This represents materials used. If 1000 amplifiers are sold and each requires a kit of components and other supplies (including packaging material) costing £35, then the cost is simply £35 000 for the period.

Shipment cost

It might cost £2000 to ship the amplifiers to a distributor.

Direct labour

If ten staff are assembling these amplifiers and each one is paid £1000 for the month (including all pension contributions, government contributions etc.), then the direct cost is £10 000.

Factory overhead

Typically, when making things, supplies and services such as electricity, cleaning supplies and services, waste removal, etc. are requred. Rent may also be being paid on the factory premises, and any factory supervisor will have to be paid. Between all of these factors, there might be a total cost of £5000 in the month.

Depreciation

The factory needs equipment for production and test purposes. This equipment might cost £120 000 and might realistically last for five years, being worthless at the end of this period. It thus loses value at a rate of £2000 per month. It is reasonable to consider this as a cost of making the amplifiers, and include it in the calculations.

The sum of these items represents the cost of actually making and shipping the product (cost of goods sold). In this case the cost of the 1000 amplifiers is £54 000, and the profit after this cost—the gross profit—is £46 000.

The picture as yet, however, remains incomplete. An amplifier maker typically needs a sales function, an R&D function and general management, including the finance function.

The R&D function might have a total staff cost in the month of £6000. R&D equipment again needs to be bought, which might cost £36 000 and be worthless after three years thus declining in value at a rate of £1000 per month. R&D consumables might cost £1000 in the month, making R&D costs £8000.

Taking together sales, general and administrative areas (SG&A), employment costs might be £16 000 to cover staff in this area. Advertising might cost £3000 for the month, and other office costs such as electricity, telecommunications and motoring expenses might cost another £2000. Depreciation on cars and office equipment might cost £1000. Total SG&A cost is thus £22 000.

Total cost in this overhead area (R&D + SG&A) is thus £30 000, which means that the profit figure is now down to £16 000 (operating profit).

Additionally, the typical business will have some bank interest expense in the period, taken arbitrarily as £1000.

Thus pretax profit is £15 000.

The tax authority will want to take a percentage—typically between 25 % and 35 %, assumed here to be one third—of this, which costs £5000. Tax will usually not have to be paid until later, but it is still a cost which will actually have to be met. After-tax profit is thus £10 000 on sales in one month of £100 000.

The company might have an approach of paying out dividends of 25 % of after-tax profits. Dividend payments in the case of public companies would typically be twice per year. A private company has freedom in when they are paid, and in this case a monthly payment of £2500 can be assumed.

A profit and loss account picture for the month is therefore

			£
sales			100 000
cost of goods sold			
materials used		35 000	
shipment costs		2 000	
direct labour costs		10 000	
factory overhead (excl. depreciation)		5 000	
depreciation of plant		2 000	
			54 000
gross profit			46 000
total overheads			
R&D staff	6 000		
R&D equipment depreciation	1 000		
R&D consumables	1 000		
R&D costs		8000	
SG&A staff	16 000		
advertising	3 000		
SG&A office costs	2 000		
SG&A depreciation	1 000		
		22 000	
			30 000
operating profit			16 000
bank interest			1 000
pretax profit			15 000
taxation			5 000
profit after tax (earnings)			10 000
dividend payment			2 500
retained profit (for the period)			7 500

This, then, is the profit and loss account, which is really a listing of all the value flows that have occurred during the month in question. Customers gave value of £100 000. Costs consisted of value bought in from suppliers of materials and services, from suppliers of labour (employees) and from the bank (fee for use of money, or interest). Depreciation charges have also been shown, which reflect the fact that equipment was losing value during the period in question. Some of the value created had to be shared with the tax authority, before finally coming up with the residual amount that represents profit or return to the owner(s) of the business.

An income statement is essentially a measurement of what has happened to a system over a period of time. The unit of measurement is money, and it shows all the dealings with the environment and changes within the system, namely deterioration of fixed assets. The income statement is shown graphically in Figure 5.1.

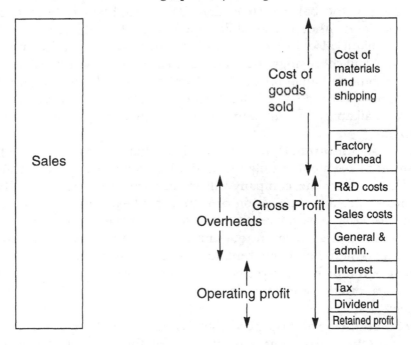

Figure 5.1   Profit and loss statement main elements

Next is the balance sheet. This is not a statement of flows or changes, but rather a snapshot of where value lies within the business (assets) and in what categories. There can also be negative values, or liabilities, representing obligations on the company's part to transfer value to entities outside the business at some time in the future.

Assets possessed by the business include fixed assets such as buildings, equipment of various types, cars, etc. These are usually stated as total purchase cost, less the total loss in value to date, i.e. less the sum of the depreciation charges. For example, the fixed assets might have cost £216 000. Based on assumptions consistent with the income statement approach, the loss in value since their purchase might have been calculated to be £66 000, and making them now worth £150 000.

A company might also have intangible assets, such as technology and other intellectual property. The treatment of intellectual property, and of intangible assets generally, in the context of balance sheets is not particularly straightforward, and these points will be discussed later. For the moment it will be assumed that there are no intangible assets on the balance sheet.

There are also assets in the form of stocks that the business possesses at the balance sheet date. These might include materials, supplies and other consumables, and might be valued at £40 000. There might also be some work in progress, i.e. partly finished product, and there might be some finished product awaiting shipment to distributors. Valuation here is a topic in itself, discussed later. For the moment, it is assumed that assembly and test teams have finished work on all amplifiers in hand, and that all finished product has been shipped.

It is an unfortunate fact of life that when a product is sent to customers, payment is only received after some weeks delay. Having sent a product, the company will invoice the customers, and these invoices—i.e. the expectation of getting paid—are naturally an asset that the company possesses, even though payment has not been received. In the case in question, there might be £100 000 of product which has been sent to customers and invoiced, but for which payment has not yet been received. Such customers, and the amounts owed, represent debtors, accounts receivable or simply receivables.

The business will also need some cash. It may be quite a modest amount, and borrowings may well exceed the amount of cash by a large figure, but the cash possessed at the balance sheet instant is an asset, and must be shown as such. It might be physical notes and coin, or might be some other easily realised source of money, usually a bank account in credit. In the current case, a figure of £5000 might be realistic.

On the other hand there is value in the business that is due to others. There are liabilities, i.e. obligations to pay money at some time in the future.

Just as customers can take some time to pay, the company might also have arrangements with suppliers to take some time to pay them. In the current example, £41 000 of items which have been received from suppliers, but for which we have not yet had to write cheques, might be assumed. These monies will have to be paid sooner or later, so this represents an obligation or liability for the company. The suppliers, and amounts owed, represent our creditors, accounts payable or simply payables.

Many companies will have short term facilities at a bank. In this case the figure outstanding is assumed to be £10 000. In some cases this would be 'netted off' against the cash position, but here a separate overdraft account is presumed, as is another account which is in credit to £5000.

The company might also have a term loan, usually secured on some of the equipment or on invoices. The importance of security arrangements is that in the event of failure of the business, the bank has the specific right to sell the equipment and to collect money from customers to ensure that it gets paid. A bank will have other rights also, but this specific right is ordinarily the most important.

In considering liabilities, a distinction is usually made between what is payable within one year and what is payable later than one year from the balance sheet date. The logic for this separation is based on the arbitrary consideration of money due in less than one year as short term debt and in any period longer than one year removed as long term debt. In this case the company might have £20 000 due within one year and £80 000 due after one year.

The final item is also a liability of sorts, in that some return to shareholders should be provided for. The return to shareholders is, however, based on sharing the profits, rather than getting invoices paid or loans paid off, so it is not quite the same as a liability to a bank or supplier. This final item on the balance sheet is termed shareholders' equity, and is calculated based on the amounts committed by shareholders to date, plus the cumulative effect of the changes in value since then, i.e. the net total of the profits less the losses and dividends since the start of the business. In this case the shareholders are assumed to have paid in £100 000 and the cumulative retained profit reported to date is £44 000.

The balance sheet would thus show:

		£
fixed assets		
cost	216 000	
less accumulated depreciation	66 000	
		150 000
current assets		
stocks	40 000	
debtors	100 000	
cash	5 000	
		145 000
total assets		295 000
current liabilities		
creditors	41 000	
bank overdraft	10 000	
current (i.e. within one year) part of long term debt	20 000	
		71 000
long term debt (less current part)		80 000
shareholders' funds		
paid in capital	100 000	
retained earnings	44 000	
		144 000
total liabilities and equity		295 000

In practice, there may be some alternative groupings of the items in the balance sheet. A common approach is to aggregate stocks and debtors and subtract creditors. This net figure is then termed working capital.

The balance sheet is useful in that it shows where value is tied up within a business, and is subject to the discipline that it must balance. This reflects the logic of double entry bookkeeping, the basic tenet of which is, paraphrasing, 'for each entry, there must be elsewhere an equal and opposite entry'. This is equivalent to stating that all value movements must be accounted for, and if so the balance sheet entries will balance. The shareholders' funds will increase by the retained profit, and this will be consistent with the changes to working capital, fixed assets, cash and debt.

The balance sheet can also be shown graphically, as in Figure 5.2. This portrayal shows that some elements in the balance sheet reflect tangible commodities, while other items represent expectations of getting paid and obligations to trade creditors and debt and equity funders of business.

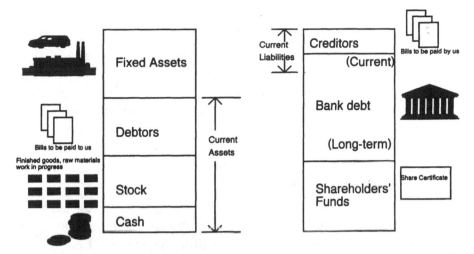

Figure 5.2   Physical links with items in the balance sheet

In addition to the profit and loss statement which records movements of value in a particular period and the balance sheet which gives us a snapshot of where value lies, the company can use one additional statement.

Cash is central to running a business and is what pays people at the end of the day. Cash generation and value creation are not, however, the same thing. The company might well be reporting profits of £10 000 monthly, but there may be a continual need to buy more equipment, or it might be necessary to allow customers some additional time in payment if the company is to get the type of business it wants. It is thus totally consistent to be reporting significant profits while seeing the bank balance dwindle.

For such reasons, many owners, banker and investors pay particular attention to the cash flow statement. This shows whether the system that is the business will not just create value, but will also create this value in the form of cash which can be made available to pay lenders and give a return to shareholders.

The items in a cash flow statement are as follows

Profit

This figure can best be stated on a profit before tax basis. The approach is to take profit as the first line of value creation and then make adjustments to see how profit translates into cash. In this case, the profit was £15 000 for the period.

Depreciation

The first adjustment to make is to add back the depreciation charge. Depreciation represents a change in value of equipment, but it does not have any impact on the cash position until the company starts buying replacement equipment. This is considered under 'purchase of fixed assets'. A depreciation charge was a part of calculating profit, and thus it needs to be added back in this context. In the example, the depreciation charge was £4000.

Movements in stocks and debtors

If the company is anticipating a big order it might, for example, have bought in a lot of materials. If sales in the month are increased, then there is usually a higher debtors figure, as customers will not immediately pay for their orders. In our example, assume that stocks have increased by £2000 and debtors by £10 000.

Increase in creditors

For a larger amount of stock, with credit terms remaining the same, the company is entitled to owe a larger amount to creditors. In this case it might be possible to have a position where creditors are owed an additional £3500.

These adjustments together may be termed working capital adjustments.

Purchases of fixed assets

Assume that the company buys a computer in the period for £1000.

Other payments

Assume that £5000 is paid in tax and £2500 in dividends in the period.

We can then show a cash flow statement as follows

	£	
sources of funds		
profit before tax	15 000	
adding back depreciation	4 000	
		19 000
uses of funds		
net change in working capital		
increase in stocks	2 000	
increases in debtors	11 000	
(less) increase in creditors	(1500)	
		11 500
fixed asset purchases		3 000
tax paid		5 000
dividends paid		2 500
total uses of funds		22 000
net cash generated		(3000)

Even with a retained profit in the period of £7500, the company has reduced its cash in the bank (or increased its overdraft) by £3000. The company made money after tax and dividends, but had an increase in trading, so that the resulting increase in stocks and debtors has more than taken up the value created in the business in the period.

These statements taken together represent a very useful model of the static and dynamic aspects of value within a business. They are key tools for management of the business, as will be discussed.

5.2 Accounting policies

Before discussing the usefulness of financial statements in designing a business, it is important to look at some of the variations in accounting policies which can be used.

In particular, in technical businesses, there is the issue of capitalisation of product development activities. The choice is either to regard development as an expense, with profits thus reduced by the amount of product development expenditure in the period in question, or as work which has a value, and which should be shown as a fixed asset to be gradually depreciated over time or over numbers of items sold.

A person with respect for technical development would probably suggest that capitalisation of product development activities is more appropriate, but the debate is more balanced than initially might be suggested. Essentially, the debate points are as follows

(i)   Writing off product development allows the business to defer tax (because taxable profits are lower in the near term, and higher later), where this policy is permissible by taxation authorities. For example, writing off software development is constrained in the United States by the requirements of FASB 52 (Financial Accounting Standards Board).

(ii) Capitalising development activity implies the creation of an intangible asset, with a corresponding increase in shareholders' funds. In practice, the valuation of this intangible asset is very subjective. One might value the development activity associated with a particular product at £200 000, based on the cost of activities, but this is a management judgment. It might only be worth £100 000, or the work might have been performed much more effectively using development tools recently introduced. The very subjective basis of valuations here, and the potential for abuse, leads most analysts, lenders and investors to just assume that R&D had been written off, and make the necessary adjustments in their own calculations. Writing off R&D is recognised as probably being excessively conservative, but at least it is a consistent policy which can be applied in comparison of companies, and one might consider some figure, such as valuation as a multiple of R&D expenditure, in judging a company, or one might look at league tables of R&D expenditure as a percentage of sales.

(iii) R&D has to be regarded as an ongoing aspect of the business. Few businesses expect to live on the basis of one hit product, and in practice the business requires ongoing R&D activity. To continually be capitalising and depreciating the R&D work associated with a series of products represents a complication which is not appropriate for a real world business.

Similar issues can arise in relation to accounting for brands. A brand such as Coca Cola clearly has value. Selling a product with a well known brand is generally easier with a higher price obtained, than in the case of an unbranded product. In either case, the brand means that a greater net profit is usually available. Money can be spent in developing the brand, as for example in expensive advertising and promotion work, and franchising out a high profile brand can give significant income. There are numerous other examples of brands having real value. Thus brands might appear on the balance sheet as intangible assets. Several scientific approaches to brand valuation have been developed, but in the end there is a measure of subjective judgment involved.

In all of these cases, there may be an attempt to match the value of the business as stated in the accounts as shareholders' funds—net tangible book value—with the value that somebody might pay for the business. In practice, however, acquirers will make their own assessment of the value of the business, and it will usually be much higher than the stated shareholders' funds, with or without such creation of intangible assets.

Another approach by which intangible assets are created is by making acquisitions. If £10 million is paid for a business with net tangible sharcholders' funds of £5 million, £5 million is being paid for assets which do not really exist, at least not according to the financial statements. Accountancy does not permit gaps and requires consistency, so either an intangible asset called goodwill must be created or shareholders' funds reduced by the amount of goodwill that would have created. Accountancy bodies are still debating policy options concerning this purchased goodwill and periods over which it might be depreciated, but most companies in the UK would not create an intangible asset and would write off goodwill, reducing share-holders' funds accordingly.

There is a role for intangible assets in certain cases, and this is really concerned with the approach adopted by comparable firms in the industry in which a company operates. To be too conservative, in not allowing any intangible assets and reducing stated shareholders' funds to the tangible figure, might penalise the company and may not be allowed for tax purposes. On the other hand, to create intangible assets to a greater extent than the other companies invites a suspicion of creativity in the presentation of results, and generally damages the credibility of the reporting firm.

Another area in which accounting policies may vary is in the treatment of work in progress and in valuation of stocks. Finished goods are usually valued at cost of materials plus a reasonable allocation of factory overhead. Following the earlier example, materials costs were £35 and the direct labour cost for each amplifier was £10, giving a total direct cost of £45. For making 1000 amplifiers a month with a factory overhead of £7000, an allocation per unit would be £7. Departing from the simple example and having some unsold amplifiers at the end of the month, these might thus be valued at £52 each.

Stocks are usually valued on the basis of cost (this can be done on a LIFO (last in, first out) or a FIFO (first in, first out) basis, but these nuances are not considered further here).

In the case of a businesses working on a long duration contract, the question arises as to how much of the expected profit on the contract can be taken in during the period in question. If a profit of £100 000 is expected on a contract of £500 000 that is 40 % finished at year end, should a profit of £40 000 be taken or should a more conservative approach be adopted? If the full £40 000 is taken and there are significant delays later on that cause the contract to make an overall loss, then results for the subsequent period are further impacted.

Leasing can also be contentious in the context of accounting policies. If a piece of equipment is leased, it is usual to focus on the monthly payment. Being realistic, however, most leases are of the finance type where in reality the asset is being bought and the lease payments represent a combination of capital repayment and interest cost. It is thus appropriate to show the lease in this light. If it is an operating lease, where a piece of equipment is being hired without any major termination cost, then just focusing on the monthly payment without any balance sheet implications is appropriate.

Property leases represent another problem area. In Britain and Ireland property leases tend to be inordinately long for the requirements of most businesses. A 35 year lease is not uncommon, and there is difficulty in getting out of such transactions. The focus in the accounts is usually on the monthly payment, and the potential risk is considered in the notes to the accounts, but this treatment in the notes does not make the liability any less than would otherwise have been the case.

Financial statements are usually also prepared on the going concern basis, where the assumption is stated or implied that the intention is that the business will continue indefinitely after the balance sheet date. Without this assumption the valuation of assets might well be very different. In the case of companies in difficulties or possibly awaiting financial support from investors, auditors may well give a qualification in their report to the effect that the going concern assumption may not be justified under certain circumstances.

An issue also underlying accounts is the making of provisions. These relate typically to stock obsolescence, bad debts etc., and one errs here on the side of prudence.

5.3 Using statements for business design

Ratio analysis

Simple ratios are very useful in computing how well a business is performing. Some of the more common ratios are as follows

Gross margin

This is the gross profit amount divided by the sales figure and gives a measure of the profitability of the business, before the costs of selling, R&D and administration are taken into account. A high gross margin is essential in a business such as an instrument maker, with continuing high costs in R&D and sales, while a low gross margin may be acceptable in the case of a manufacturer producing a standard basic product in volume for large customers, where R&D and sales costs are proportionately much less.

Net margin

This is the profit reported after taxes, divided by sales, and is really a measure of the earnings capacity of the business. The acceptable figure here is really related to capital employed in the business. If a lot is invested in the business, such as in a heavy manufacturing operation or a business which has to carry large stocks, then a net margin of the order of ten per cent might be the minimum demanded. If the investment requirement is much less, as in the case of a well managed distributor or a supermarket, then a very successful business can have a net margin of perhaps two per cent.

Return on capital employed

To an extent this is the key measure of business profitability and there are several variations. The return on shareholders' funds or return on equity gives the return that an investor in the business is making. This is central to the issue of being in business in the first place; the company must be consistently making a return that is better than putting the money on deposit in a bank. Losses or below market returns can be tolerated for a short period of startup or repositioning the business, but the expectation has to be for good long term satisfactory returns. The return on capital employed may be considered to be the operating profit divided by the sum of shareholders' funds and the bank funding in the business. This figure gives an overall measure of return on the funds employed (equity and debt) in the business, and is another useful score of information on how the company is doing. In practice, return on capital employed of about 15 % to 20 % might be sought in a manufacturing business, and perhaps 20 % to 30 % in a distribution business. For a particularly well placed company with business under very good conditions, the return on capital employed may exceed 100 %.

Fixed asset turnover

This figure is given by sales divided by fixed assets, and is a measure of the amount of fixed assets used for the current sales level. Once again, the acceptable figure varies with the business. A telephone operating business might need fixed assets which are typically one and a half to two times annual sales. For a distribution business, buildings, office equipment and fittings needed might together amount to more like ten per cent of annual sales.

Working capital ratios

Important measures in most businesses are how quickly debts are collected, how quickly creditors are paid and how quickly stocks are turned. Typically, these are measured in terms of days, or times per year in the case of stock.

For example, if annual sales are six times the figure for debtors, then the company has a debtors' figure of two months or, as more usually stated, 60 days. The shorter this period, naturally, the better. If business can be done on a cash basis like a supermarket, then the company can have zero debtor days. Most businesses in technology areas are not quite so fortunate! Debtor days need to be minimised in a way that is consistent with getting the business in the first place and not having to offer excessive discounts to reduce the figure.

Likewise, if annual purchases are twelve times the figure for creditors, then the creditors' figure is one month or, as more usually stated, 30 days. This figure needs to be as high as possible on an agreed basis, as taking excess credit beyond the terms agreed leads to poor relations with suppliers and to concerns about the solvency of the business. A reduction in this figure might be accepted if good payment discounts can be obtained.

Stock/sales represents a useful figure as something to be minimised subject to the constraint that adequate stock levels must be kept to run the business, and that stock should be bought in economic quantities rather than placing large numbers of small orders, possibly involving loss of quantity discounts and additional shipment costs. A large manufacturer may be able to virtually eliminate stock from the manufacturing process by insisting on a just in time delivery schedule from suppliers, but not all businesses have the strength to demand such performance from suppliers.

The overall figure of working capital divided by sales gives an indication of what is required by way of funding to achieve a near term increase in sales, assuming that the fixed asset requirement is unaltered. A figure of 0.2 means that for each extra £100 000 in annual sales, extra working capital funding of £20 000 needs to be found.

Gearing/leverage ratios

Most businesses have some element of debt funding. In some cases, such as where the business has a very limited equity base, debt is taken on as a necessity if the company is to have any chance of growing. In other cases, prudent use of debt may be desirable in order to achieve a better return on shareholders' funds. The characteristic of debt funding is that it involves a return not directly dependent on the success of the business. Assuming a business with shareholders' funds of £100 000 and debt of £100 000 with operating profits of £50 000, return on capital is thus 25 %. Debt demands interest payments of perhaps ten per cent, or £10 000 in this case, leaving £40,000 profit (pretax) for shareholders, or a 40 % return on shareholders' funds. If all the £200,000 had been supplied from shareholders, then there would only be a 25 % return. On the other hand if operating profits were to drop to £15 000, the lender would still have to be paid £10 000, giving a return on shareholders' funds of just £5000 on £100 000, or five per cent. If the business was funded fully with equity in this case, the return would be 7.5 %.

Hence debt allows a company to leverage or gear its return; interest payments on debt are usually also tax-deductible. There are obviously limits on the extent to which debt should be taken on in a business. Interest on debt needs to be paid continually, whereas shareholders may be asked to forego dividends, and lenders will impose conditions on the business, in order to protect their loan. Debt may also be totally inappropriate in the case of early stage businesses, in particular businesses which are not yet ready to achieve profitable levels of sales. In some cases lenders may seek personal guarantees from owners, which is a form of funding that has neither the attractive character- istics of conventional debt nor of equity funding.

Such ratios come in various definitions. A ratio of debt to net tangible shareholders' funds may be a useful measure. This might be 50 % in the case of a conventional business, or might be very much larger in the case of highly leveraged transactions (where share- holders' funds may be negative making the figures meaningless). This type of transaction is often encountered in management buyouts, as discussed later.

These statements, and the ratio analysis that can be undertaken, provide tools for analysing the business and seeing which areas to emphasise during business design. The goal is to get the maximum long term return on shareholders' funds, but this is subject to a number of constraints imposed by both the environment and the entrepreneur's own desires. To achieve this goal the company needs maximum sales and maximum profitability, but profits are a function of what the marketplace will allow. The sales figure is governed by what the market will buy from the company, and also by the company's ability to afford the equipment needed, and particularly the working capital associated with having to buy-in extra stocks and endure the extra debtors figure. An increase in short term profitability can also be achieved by reducing R&D or sales expenditure, but at significant long term cost. Gearing is limited in practice by what the bank will permit, but also by judgment of the amount of risk to be taken on at any given time.

5.4 Illustrative financial statements

Some illustrative typical financial statements for types of businesses other than the small manufacturing company already discussed and relevant to the technical entrepreneur, with comments on them, are given below. These have been simplified to show the key issues involved.

5.4.1 The software consultancy

A successful, five person consultancy operated as a limited company, with two principals having shares in the business.

profit and loss statement for year ended 31 December 199X

		£
income from consultancy		600 000
costs of work performed		
staff	330 000	
depreciation of equipment	20 000	
materials	30 000	
expenses	20 000	
		400 000
gross profit		200 000
sales and administrative costs		90 000
operating profit		110 000
interest income		10 000
net profit		120 000
taxation		40 000

		£
PAT		80 000
dividend		50 000
retained profit in period.		30 000

balance sheet at 31 December 199X

fixed assets		
at cost	100 000	
less accumulated depreciation	(40 000)	
		60 000

current assets		
debtors	100 000	
work in progress (net of progress invoices)	50 000	
stocks	10 000	
cash	60 000	
		220 000
total assets		280 000

current liabilities		
trade creditors, miscellaneous	5000	
taxation	40 000	
		45 000

shareholders' funds		
paid in capital	10 000	
profit and loss account	225 000	235 000
total liabilities and shareholders' funds		280 000

This business is representative of a well run consultancy, where the business pays out a high proportion of profits to the shareholders by way of dividends, and yet retains a good measure of cash and has no debt. Having such a structure is probably appropriate given the somewhat lumpy nature of consultancy business. Being well funded to cover the inevitable lean period is also appropriate, as banks will be less than comfortable with lending in such situations and there will be limited assets to be taken as security. In the case of consultancies, there is usually also the issue of long term contracts and how to recognise the income from these. If the contract looks on target and there are clear milestones, the profit can be credited uniformly over the project, but a more conservative approach might be suggested if the project is complex or novel. Mention is also made of progress payments, and it very desirable to have such an arrangement in place in the context of a long job, ideally with payments matching the percentage completion of the job, but at least covering the cash expended at each stage.

5.4.2 The distributor

This company is a distributor of electronic components to large customers, to whom it sells in volume. The profit and loss statement might thus be as follows.

profit and loss statement for year ended 31 December 199X

	£	
sales	12 000 000	
costs of sales		
product supplied by manufacturers	11 000 000	
gross profit	1 000 000	
sales and administrative costs		600 000
operating profit	400 000	
interest expense	100 000	
net profit	300 000	
taxation	100 000	
PAT	200 000	
dividend	50 000	
retained profit in period	150 000	

balance sheet at 31 December 199X

fixed assets

at cost	500 000	
less accumulated depreciation	(80 000)	
		420 000
current assets		
debtors	1 500 000	
stocks	500 000	
cash	10 000	
		2 010 000
total assets		2 430 000
current liabilities		
trade creditors, miscellaneous	700 000	
bank overdraft	100 000	
current part of term facility	200 000	
taxation	100 000	
		1 100 000
long term debt (less current part)		500 000
shareholders' funds		
paid in capital	100 000	
profit and loss account	730 000	
		830 000
total liabilities and shareholders' funds		2 430 000

This represents a distribution business which operates on very small gross margins, (8.3 %) and has net margins after tax of only £200 000 divided by £12 000 000, or 1.67 %. One might question whether it was attractive to do business at such levels of profitability. The answer relates to the return on capital employed. The shareholders' funds in this example are £830 000, so obtaining an after tax profit of £200 000, or 24 %, even after the use of significant leverage, may be acceptable.

Another observation is that businesses with such low margins are very vulnerable to failure of a major customer. It only takes the failure of one customer in such a scenario to wipe out the profit for the year. If 60 days credit is assumed to be taken by customers, then to run up £300 000 (the pretax equivalent of £200 000 net profit) requires an annual turnover of £1.8 million. If the company is doing business on normal credit terms with one customer providing 15 % of business and this customer fails, the company loses its annual profit.

This business is also carrying significant debt, i.e. carrying £800 000 against £830 000 in shareholders' funds, or 96 % gearing. Banks might find this acceptable, especially if they have confidence in the individuals involved, if there is good cash-flow from the business and they are well secured on receivables, but such a situation might make the owners less than comfortable if trading goes into a difficult patch. In businesses of this type suppliers may also take an interest in the financial statements of their customers. They may have retention of title clauses, saying that they retain title to goods until they have been paid in full, and these clauses may conflict with clauses required by banks.

5.5 The finance needs of a business

A business obviously needs funding in its early stages if it is hoping to undertake a program of product development before it succeeds in selling product, or if it is going to incur losses.

Looking at the balance sheet of a business, it can be seen that a business needs assets, and these assets have to be financed. If a business has annual sales of £1 million, then it might typically need buildings, plant and equipment of £200 000 to £500 000 in total value—say perhaps £250 000. If the debtors pay in 60 days and the company needs stocks for 30 days—making conservative assumptions —it might need an additional £200 000. Thus, assets of £450 000 need to be funded.

Creditors will help to a small extent—perhaps £30 000—which leaves £420 000 to be found. The wisdom of banks and the company's own caution would suggest that this is funded with at least £250 000 to £350 000 in equity, or between 25 % and 35 % of annual sales revenues.

This is all very well if the entrepreneur has a large inheritance, and the business can just start without initial losses. In practice, most businesses have to start in a small fashion with proprietors' equity and with bank borrowings in ratios which might seem excessive in the case of a larger business, but for which personal guarantees may have been given, which is really commitment of 'equity in disguise' by the founder. By excessive work and with limited salary taken from the business, gradually retained earnings bring shareholders' funds up to the level where they can support these levels of annual turnover.

It is clear that there is great advantage in reducing the assets required to carry on a certain level of sales. If manufacture can be subcontracted, the company may well be able to do without a large part of the fixed asset requirement, and might also do somewhat better by having lower stock requirements and getting good credit terms from the subcontractor. On the other hand the company may lose out on margins, quality may be less obviously under its control and, in busy times, responsiveness of the subcontractor may suffer. If customers could be persuaded to pay on time, this is perhaps the most important means of minimising our funding requirement. Factoring customers' invoices (essentially selling the invoices to a factoring company for an up front payment) could also be considered, but this is really taking on debt under a different guise. It might be worth offering significant incentives for up front payment, or the company might seek to redesign the business approach to work more with customers for whom prompt payment—e.g. by letter of credit—is the norm.

Recognising that the profit and loss statement represents a flow in a period, and that the balance sheet represents a snapshot of the financial condition at a given moment, provides the basis for an equation which can give us the natural growth rate for the business.

Defining terms:

S   —sales per unit of time t
OP—operating profit margin
G   —maximum permitted gearing ratio, debt (DT)/shareholders' funds (SF)

F —fixed asset/sales ratio
W —working capital/sales ratio
I   —interest rate per unit of time
T —tax rate
D —dividend payout ratio
RP —retained profit per unit of time

The profit and loss statement is effectively

$$RP = (OP \times S - DT \times I) \times R$$

where R is the retained profit ratio

$$R = (1 - T) \times (1 - D)$$

and RP represents the rate of change of shareholders' funds.
From the balance sheet

$$S \times (F + W) = DT + SF$$

Operating with a fixed gearing ratio gives

$$S \times (F + W) = SF \times (1 + G)$$

or

$$S = SF \times \frac{(1 + G)}{(F + W)} = K_1 \times SF$$

where

$$K_1 = \frac{(1 + G)}{(F + W)}$$

DT is defined as $G \times SF$, thus $DT = G \times (S/K_1)$

$$RP = \frac{d(SF)}{dt} = \left(OP \times S - I \times (G/K_1) \times S\right) \times R = K_2 \times S$$

where $K_2 = R \times \left(OP - I \times (G/K_1)\right)$
or

$$\frac{d(SF)}{dt} = K_2 \times S = K_1 \times K_2 \times SF$$

This is a first-order differential equation of the form $dy/dt = ay$, for which the solution is

$$x(t) = Ke^{at}$$

Thus

$$SF(t) = SF(0) \times \exp{(K_1 \times K_2 \times t)}$$

This relationship suggests that the natural growth rate is determined by these factors. Assuming a set of realistic parameters such as

$$T = 0.33$$
$$D = 0.25$$
$$F = 0.25$$
$$W = 0.2$$
$$I = 0.10$$
$$OP = 0.10$$
$$G = 1$$

then

$$R = 0.5025$$
$$K_1 = 4.44$$
$$K_2 = 0.0389$$

Thus the exponent is 0.173, and the maximum growth rate for the business in a self-funding fashion is 19 % annually. The analysis has naturally involved a number of simplifying assumptions, perhaps the main one being that business operates as a continuum, but it does imply that if a company is seeking to grow at a faster rate than this figure, then it will need to take in further equity. If the capital intensity of the business can be reduced by reducing F and W, then a faster rate can be obtained, as naturally is the case if cash is conserved within the business by low dividend payout ratios. If the D figure is reduced to 0 and F + W from 0.45 to 0.3, then the natural, or self-funding growth rate of the business increases to 46 %. Looking at the effect of gearing, if G = 0 in the original example, the growth rate is now 12 %.

The main stages of development of a business at which equity funding is required are thus

- at the early stage, where product development is being undertaken prior to sales being achieved
- funding early stage trading losses, where profitability has not yet been achieved
- in funding growth of a business at a faster rate than permitted by its retained earnings (i.e. faster than the natural growth rate figure derived approximately above).

Very loosely, the initial stage of R&D funding with no sales can be described as seed capital funding, the early stage period where trading activities are not yet at levels that will make the company profitable, as venture capital funding, and funding in support of growth of an already profitable business as development capital funding. The term venture capital, or more generally private equity, can be applied to all funding stages in unquoted (i.e. not publicly traded) companies.

In addition, debt funding may be appropriate. Debt funding for a business is typically tied to certain assets of the business on which security can be taken by the lender. The most obvious assets are the debtor invoices which, if they are for proven products being sold to large and reputable clients, represent a valuable asset. If the company is selling untried products, or selling services, or selling to a less financially strong customer base, then the value of the invoices may be more open to dispute, and their value as an asset be more questionable. In this case, raising funding based on the value of the invoices may be more difficult.

Debt funding related to fixed assets is also the norm. The simplest case is when buying a factory or office premises. It is relatively easy to raise debt finance secured on the land and buildings, not dissimilar to taking out a mortgage to buy a residence. If the company has equipment, then securing debt funding based on its value is usually feasible. The readiness of lenders to provide funding is based on the likely resale value of the equipment. In some cases, such as in the printing industry or in the case of vehicles, there is usually a realistic realisable value on second hand plant, but getting debt funding based on specialised equipment may be difficult. Lenders will, of course, also require that assets are well maintained and insured. The type of business and the likely characteristics of the assets will be well known to lenders, who may well be more prepared to lend money to some types of businesses than to others. There is also judgment of the personalities involved by lenders, who are naturally happier to lend money to firms owned by individuals with particularly good commercial track records.

Being able to get security is one aspect of the picture from the lender viewpoint. Having some comfort that the business will be able to generate cash to pay interest and reduce the balance outstanding over time is also required. Lenders will usually require some indication that the business is likely to be profitable, and that profits are some significant multiple of the interest charges. Projections of profitability and of cash flow are thus typically sought by lenders.

Funding the business is now considered in more detail in Chapter 6.

Chapter 6
Building the equity base of the business

Businesses usually require significant equity capital, particularly if they are funding product development or marketing programs or if they are expanding rapidly. Firms in the venture capital business, wealthy individuals, and indeed stock exchanges, can be appropriate sources of such funding.

Before seeking to raise equity capital from such sources, it is appropriate to look at other means of building up the equity base of the business.

6.1 The soft start-up and other limited funding situations

The term 'soft start-up' can be applied to a situation where the individual or team initially undertakes consultancy work. This work is usually undemanding in terms of fixed assets, can be profitable and, with good clients and well designed progress payment arrangements, it can be quite cash generative. Operating a consultancy can usually also allow the promoters to do development work, perhaps on an after hours basis, so that product development can be undertaken at very low apparent cost. An even better approach is if the right to retain the intellectual property associated with custom design work done for some clients can be negotiated; this may form the basis for products. Clients may be very happy to fund development of a specific product, perhaps with price concessions, with the understanding that it will not be released for general sale until after the expiry of a six month period from completion to their satisfaction.

Persons engaged in academic work may also have the opportunity for low cost product development, perhaps with payment of a subsequent royalty to the institution in question. There may also be government or international (e.g. EU) research contracts available, which can lead indirectly to development of product opportunities.

The firm with innovative products or technologies can also get significant funding from selling technology licences to others. For instance, a US company may recognise that it will have limited resources to address the European market and may thus be happy to sell rights in advance to a European company, or vice versa. Cash received from sale of these rights can usually be recorded as clear profit and can thus fund the expansion of the business.

In addition, many countries and regions maintain schemes of grant assistance to help firms, particularly at the early stage. If such grants are used wisely in association with some of the other means of building a company's equity base, then progress can be faster.

Many service businesses are not particularly cash hungry in their early days, and funding them up to a level of profitability and positive cash flow may be possible from the resources of the promoter, assisted as required by family and friends. Even with manufacturing operations, purchase of second hand plant and renting cheap premises can allow a low cost commencement of operations.

A summary of some of the approaches to starting a business with limited resources is given in BHIDE, 1992.

6.2 Raising venture capital

There are, however, some situations where this self funding or bootstrap approach does not work, and it is necessary to secure equity capital from outside sources. It may be that to take advantage of an opportunity the company will need to move much more quickly than is possible if the founders rely on their own resources. Or it may be that an extensive R&D programme is required if the company is to be taken seriously, or a dedicated facility may be required if customers are to even consider giving it business.

This capital from outside sources may come from taking in another active shareholder, or there may be individuals prepared to invest, particularly if tax relief is available. The main focus here is, however, on finding professional investors, termed venture capitalists.

Venture capital is usually applied to all forms of investment in companies without a stock exchange listing. It can sometimes be segmented into the categories identified above, i.e.

- seed capital, to fund start-up of a business with initial R&D and market development work
- early stage capital, typically describing funding at the stage where the business is beginning to record some sales, but where it is still reporting losses or very small profit figures
- development capital, where the business is reporting some profits, but where these profits are inadequate to fund the rate of growth desired.

A significant part of the venture capital investment is also committed as replacement capital, either to allow the buyout of an existing shareholder in a business or to undertake management buyout (MBO) or management buy-in (MBI) transactions. MBO and MBI transactions will be discussed in subsequent sections.

In raising equity funding from the venture capital industry, part of the company is being sold. As in selling anything, the motivations of the buyer need to be understood, and the proposition presented very professionally.

The venture capital investor is usually interested in making significant capital gain on his investment after a period of five to seven years. Most venture capitalists will seek some modest dividend income (or running yield) while they have the investment in the business, but the primary objective is to secure this capital gain.

Investment opportunities are considered with this objective in mind, and the five to seven year period is driven by the fact that most venture investors operate funds of ten years' duration. It is usual for most of the investment activity to be undertaken in the first three years of the lifetime of the fund, and the final two years of the fund life are usually spent in selling investments in order to be able to return cash to the fund investors at the end of the ten year period, or during whatever permitted winding up period is allowed thereafter. Even in the case of investors who invest corporate funds and who are not subject to a ten year fund life, the business is looked at in the context of it being able to create significant value in this timeframe.

Before investing, the investor will thus typically be focused on the exit; it is often much easier for the investor to commit funds than to sell the investment later on. An investor has to form views as to the likely development of the industry sector—will it be possible to achieve a satisfactory stock market listing for the business, or will it be possible to find an industrial or trade buyer?

Parenthetically, one ought to note that the process usually leads either to sale of the business or to its flotation on a stock market. If neither of these outcomes fit in with the objectives, then venture capital is unlikely to be a relevant funding mechanism. Only in rare cases will it be possible for the company to fund the buyout of the venture investor's interest, or will another investor be found to take over the interest.

A typical venture fund might have £30 million to invest. The institutions providing the funds will probably allow the management around 2.5 % of this for fund management purposes. Good venture capital executives have high earnings expectations; additional costs will include funding a significant amount of travel, as well as maintaining offices in a central (i.e. high rent) location and administrative and reporting overheads. A venture fund of this size can typically thus afford a front line team of perhaps five executives. Between scouting out new opportunities and staying in touch with current investments, a typical executive will be limited to perhaps four investments, or twenty investments for the overall team. This works out at an average commitment per company of £1.5 million.

There are, of course, some overly simple assumptions here, but the example serves to illustrate why venture funds show a bias to larger investments. In some cases venture investors can manage to have more investments per staff member, perhaps restricting their activities to a very narrow sectoral spread or to a narrow geographic region. Some seed capital funds have also been very successful, investing modest amounts of the order of £50 000 to £100 000 per project, with the secret for their success being detailed knowledge of the companies and sectors involved, as well as an ability to work closely with the founders.

The point, however, is that most venture investors are not interested in commitment of, say, £250 000 to fund modest growth of a business, and prefer to commit a much larger amount of money, perhaps in several stages, to a company with larger growth objectives.

The inability of the formal venture capital industry to look at smaller projects, or those in the early stages of development, has created the opportunity for 'business angels'. Good wealth generation opportunities in recent years have created a relatively large group of people, many with experience of technology businesses, who may be prepared to invest sums up to around £100 000, and in many cases far larger, in early stage businesses showing promise. Tax relief, for example under the EIS (Enterprise Investment Scheme) in the UK and similar measures in other countries, may also be of value. Such angels are often able to contribute considerable wisdom through playing, perhaps, a semi-executive role with the companies in which they invest.

The venture capital business is about measured risk taking. In a portfolio of ten venture capital investments it is not unusual to find two big winners, perhaps giving a gain when sold of five to 20 times the initial commitment. There will also be a number—hopefully very small—of total loss situations, and some modest growth businesses giving a return that does not dramatically exceed that obtained by leaving the funds on deposit in a bank. If there are the big winners, these can more than compensate for the losers and the laggards.

However, while there is fatalism associated with the expectation that some projects will be duds, there is usually intense homework, or due diligence, involved in making sure that any individual deal represents a sound investment. The risk factors will be acknowledged, but if the perceived probability of large success outweighs these risks, then the deal gets done.

In assessing the opportunity, the venture investor will look initially, as previously noted, at the exit opportunities. There will be a focus on the market—will the market be receptive to the product, will it have sustained competitive advantage relative to existing and planned product offerings, will the company have to grow very quickly in order to match industry growth? The usual questions associated with an industry analysis will also be undertaken.

Promoters of the company will be studied. Can they be worked with? Are they good business partners? Is the team complete in key technical and marketing areas, or can good people be recruited? (The finance role is also key, but this is a less industry specific area, and thus recruitment of good people is easier.) There is the adage that the combination of grade A management with a grade B market opportunity is to be preferred to the combination of grade B management with a grade A market opportunity!

The logic here is that management is largely a constant factor (particularly if the investor is going in as minority partner and is not seeking to have the power to change management) whereas the marketplace is ever changing and good management is needed to guide the business through evolving market conditions. On the other hand, if there is a particularly interesting market opportunity, the venture investor may be prepared to make the commitment of time and effort necessary to build a strong management team to give the venture a higher probability of success.

A practice common in the venture capital industry is syndication, particularly for the larger investments. Sometimes, the proposed business may be too large for a venture investor to take on alone, or sometimes another fund may have particularly strong connections in a relevant geographical area. If the world market is being addressed from a European base, then it makes sense to bring in partner investors from, perhaps, North America and Asia which have particular skills and contacts with relevant customer groups in these areas.

Venture capital funding of a growing business may be undertaken in stages. The initial commitment might be for £500 000 at, say, £1.00 per share, with subsequent stages involving, usually, larger amounts (to reflect the increased funding needs of the company at later stages of product or market development or sales growth) and higher share prices, reflecting the growth in perceived value of the business as it achieves additional development milestones or sales volumes and profits.

The venture investor will also be very selective. Any one venture fund will usually receive several tens or hundreds of propositions annually, depending on its declared areas of specialisation and selling and promotional efforts with accountants, bankers, solicitors and others perceived as intermediaries, as well as from direct advertising. The number of transactions actually undertaken will be perhaps five or six new situations annually for a fund of about £30 million. Part of the selectivity comes from the high time commitment necessary to research and negotiate possible deals, as well as the significant time commitment associated with supporting the development of existing investee companies, arranging additional stages of funding for them, etc.

Hambrecht and Quist, a leading US venture capital investor, lists five major attributes which it seeks prior to investment in companies

1 *Timing ahead of the curve* We like to be early in a market. We seek out entrepreneurs who in turn are seeking out new opportunities, new ways of doing business and new markets. We think the greatest opportunities are in areas that are not currently well understood, such as markets for which there is no real market research because the market is not well defined. These new markets could be an outgrowth of discontinuities or changes in technology, in governmental policy, in economic forces, or the like.

2 *Growth market* We look for markets that barely exist today but which are likely to grow rapidly to significant size.

3 *Compelling product* We look for compelling products or unique services. We want to understand the buying motivation and why the product creates a compelling purchase decision in the company's target market.

4 *Sustainable competitive position* We look for a sustainable competitive edge. We assume that the companies we back will be successful with their initial products. Beyond that, we want to understand why they will be able to maintain their success through continued stages of corporate growth.

5 *Talented people* We look for people who have been successful either at prior startup companies or as part of larger companies. We look for no universal career path: the people we have worked with successfully have come from highly diverse backgrounds. A common characteristic is success in previous endeavours and commitment to the next opportunity.

Each venture investor will have their own preferences, perhaps not set out quite as clearly as those in the list above. Many will have a preference for later-stage transactions, and many will not be so technology oriented in outlook and capabilities. Some investors will place very high reliance on their perception of the capabilities of the managers involved, giving much less attention to industry analyses, while others will focus on the industry initially, and then look at management capabilities, and whether they may need to be augmented in the light of the perceived opportunity. In exceptional cases there will be efforts at team-building by the investor.

Getting a match between management capabilities and interest of the venture capitalist is a key aspect of understanding the buyer behaviour issue. In general, in Europe, the focus of venture capital interest has been on larger development capital transactions in basic business areas rather than in technology areas, but possibly with a promise of a greater return to interest in technology investing as the established areas become more competitive leading to higher deal pricing and consequent pressure on returns. A perspective from the British Venture Capital Association is given in ARUNDALE, 1995.

Venture capital statistics may be somewhat inexact in nature due to varying definitions of stage of investment, and due to the fact that almost all venture capital deals are custom designed with various provisions to reflect the exact objectives of the parties involved. In Europe, total investment in 1994 in venture capital was reported as ECU5.4 billion, (approximately £4 billion) with £1.77 billion of this in the UK. Of the European investment figure, five per cent was in start-up situations and 0.7 % in seed stage transactions, with the corresponding UK figures being 2.4 % for start-up investment and 0.2 % for seed stage. Looking at percentage of transactions undertaken, due to the smaller amounts committed to early stage transactions, the European figure is 13.1 % for start-up investments and 3.8 % for seed stage, with the UK figures being 5.9 % and 1.6 % respectively (source evca/Ernst and Young, 1995).

In Europe, the majority of venture capital funding is put into management buyout and management buy-in transactions (44.1 % in Europe, 64.8 % of investment in the UK in 1994), and other later stage businesses, with relatively little focus on technology businesses and earlier stage situations. In the United States, there has been a considerable focus by the venture capital industry on technology businesses, with a key role being played in development of many of the major companies which currently dominate the electronics industry. The relative difficulty in getting funding for high-technology companies in Europe is ascribed to

- lower levels of willingness on the part of potential entrepreneurs to move out from academia and from successful larger companies
- greater difficulty in launching a worldwide product from a European base. In Europe there is not the large, homogenous, innovation receptive marketplace which characterises the United States. Also in Europe, there are fewer true centres of excellence— in a commercial sense—than can be found in regions in the United States
- a less well developed exit mechanism. The mid-1990s has seen the London Stock Exchange become a satisfactory home for many technology companies, but this market is dwarfed by the investor interest, broker coverage, liquidity and fund raising capability available, particularly in the case of technology companies, on the NASDAQ (National Association of Securities Dealers automated quotation system) market in the United States. This model has prompted plans to set up the EASDAQ market in Europe by the European Association of Securities Dealers in the late 1990s. The shortage of exit mechanisms for technology companies has been described as the *Achilles' Heel of Europe* (*Financial Times*, 5 March 1996). The absence of well-developed exit mechanisms means that companies are more likely to be sold outright at a relatively early stage in their potential development, whereas availability of exit mechanisms can allow the venture investors' interest and some of the management shareholding to be replaced by new shareholders in the context of the public listing, while the company retains its independent dynamic.

The environment for the entrepreneur, particularly the technology entrepreneur, in Europe may thus not be as attractive as in the United States. On the other hand, competition for funding from the venture capital houses which do specialise in technology investment, possibly linked in syndication with US investors, may be less acute than can be encountered in the US. In some areas such as in medical devices, the charge is that delays in product approvals in the US may allow particular opportunity to European firms.

Given some understanding of the buyer behaviour of the venture investor in the context of seeking to sell shares to raise equity, the company seeking funds then has to look at how to meet the criteria. It needs to communicate the details of its business opportunity in a business plan document, and needs to be prepared for a large number of meetings and presentations, some of which will be for the purpose of communicating information necessary to assist the potential investor with his research on the market, and some of which will be for the purpose of helping the investor get to know the team with whom he or she will be working in partnership for a considerable period.

The approach can often come in the other direction. Venture investors are continually bemoaning the shortage of good investment opportunities. If a business has received some measure of publicity, then it may well receive calls from potential investors, This usually makes the capital raising process somewhat easier, as such an investor is obviously displaying initial enthusiasm for the business and has made the first move.

6.3 The business plan

Business plans are prepared for numerous purposes. The most important purpose is as a tool for management, for working out the strategies and determining the resources necessary for development of the business opportunity. A well written document in pursuit of this objective is also a powerful selling tool when it comes to raising equity (or for that matter, debt also) funding.

The business plan presentation usually follows a logical sequence.

Initially, there is an executive summary, perhaps one or two pages in length. Reading this allows the investor to decide whether or not to take the investment opportunity seriously, so good presentation is essential.

It is then necessary to present an industry analysis, outlining overall industry characteristics such as growth rates, segmentation and niches, and national or regional issues. A discussion of technology is appropriate, as is a presentation on other companies addressing the industry and their characteristics and apparent strategies. Characteristics of key customers of the industry can be communicated, as can information about key suppliers to the industry. In the case of an emerging industry details here will be very sketchy, but having some views as to the likely industry characteristics is important.

Then comes a presentation on the company's goals in this industry, why these goals are realistic and achievable, and the strategy for achieving these goals. Marketing and product strategy must be outlined, along with areas of competitive advantage, how these areas are going to be maintained and how others will develop over time. The company needs to outline its production/operations approach, and how it is going to source suitably skilled staff.

Presentation of strategy needs to have a strong quantitative aspect, but formal financial projections are usually presented in the next section. The requirement is a summary of expected sales and profitability in each year and funding levels required, and then detailed projections of profit and loss, balance sheet and cash flow statements for a three to five year period, perhaps with a monthly analysis for the first year or two years.

A section outlining the CVs of founders and other key members of the management team, with particular emphasis on areas of proven success (in any area) and on industry knowledge and experience directly relevant to the business in question, might be included next.

A summary is then appropriate, outlining again the key strengths of the opportunity and again stating the amount of funding sought.

Appendices giving further market information, product brochures and copies of relevant articles in trade publications may also be useful.

Extracts from a sample business plan for a notional venture follow. The plan is based on the case of a fictional company Digital Instruments Limited, which is addressing the likely future market for test equipment for the universal mobile telecommunications system (UMTS) systems now at the research stage. A technology not yet in current use is chosen for this illustration, and the plan is dated 2005, in order to avoid any implied comments on vendors of current products addressing mobile radio testing markets.

Business plan for
Digital Measurements Limited

[Disclaimer Notice]

Executive summary

Digital Measurements Ltd (DM) has identified an interesting opportunity in the area of telecommunications measuring equipment.

UMTS systems are currently being deployed extensively throughout Europe, with penetration expected to rise from current (2005) levels of one per cent of population to 20 % of population by 2010. These systems allow transmission of data at rates of 2 Mbit/s to mobile users, and the growth rates projected are based on continuing user demand for mobile access to video on demand and videotelephony services, fast database access and remote computer operation with full graphical user interface (GUI) capabilities. DM sees a clear need for test equipment for base stations and for mobile units compatible with the new protocols being employed. Independent estimates for this market in Europe in 2010 suggest a value of £200 million, divided equally between network operators and equipment resellers. Manufacturers of test equipment for established narrowband group system mobile (GSM) (13 kbit/s) products are addressing this market, but DM believes that use of a new architecture in product design will give significant advantages to the users.

DM has completed development of a range of products, and is understood to be well placed to gain patent protection on key technology aspects. The company maintains an ongoing development program, and has a number of product attachments and enhancement options under development which will maintain its technical competitive advantage.

DM currently sells directly in the United Kingdom, and has identified key distribution partners to give coverage throughout the rest of the European Union (EU).

Manufacturing arrangements have been negotiated with a leading contract assembler and with a leading field maintenance organisation operating throughout Europe.

DM expects to gain 30 % market share in the EU by 2010, resulting in a business with annual revenues of £60 million and pretax profits of £6 million. To achieve these levels of performance, DM wishes to secure equity funding of £2 million now, with a further amount of £2 million required two years from now.

The DM team is led by a chief executive with extensive industry experience, having been instrumental in bringing a vendor of narrowband digital mobile (GSM) test equipment to market leadership in Europe. The technical director and the core technical team is already employed with DM and persons to fill the key roles of marketing director and finance director have been identified and are committed to joining DM at the conclusion of the current funding.

The directors believe that DM represents a high quality investment situation, with the opportunity of building a valuable business having good market share in a fast growing niche market.

Business Plan for
Digital Measurements Limited

Introduction

Digital Measurements was founded in 2003 with the objective of establishing a valuable business in the area of test equipment for the developing UMTS (universal mobile telecommunications system) service, launched recently throughout Europe.

The market opportunity

UMTS allows fast data transfer at rates of approximately 2 Mbit/s to mobile users. This data rate allows for videoconferencing, provides the ability to access video on demand services and allows for very fast file transfer to and from mobile computers. UMTS also caters for voice service and is widely seen as the successor to the GSM system used for voice and narrowband data transmission since 1993, which has established penetration levels of 40 % of the population in many EU countries.

The UMTS infrastructure is currently under construction across Europe, and it is expected that 95 % coverage (in terms of population), and coverage along all major air, road, rail (including underground/tunnel systems) and sea routes, will be in place by 2010, with expected penetration levels of 20 % of the population.

DM has identified the market for specialised test equipment for UMTS as being of particular interest. Currently, testing is usually undertaken in-effectively using combinations of equipment designed for standard radiofrequency testing in association with standard digital communications test equipment.

DM has commissioned independent market studies which suggest that the market for specialised equipment for testing of UMTS systems will be worth £200 million annually by 2010 in Europe. Copies of these studies are available to interested potential investors. The markets in the United States and in Japan use technology incompatible with European standards, and these markets will not be addressed by DM. The European standard is likely to be adopted in other regions of the world, which may represent an overall market opportunity for DM greater than the £200 million figure forecast.

The market is essentially segmented into equipment for base station testing and for testing of portable equipment, and is likely to divide equally between these two segments.

Base station test equipment is usually sold to network operators, with manufacturers of such equipment also buying product.

Key requirements for selling to network operators include

- ease of use of product, and light weight and robust design. Product must thus be well designed in terms of data capture and analysis software, with good communications capability, to allow analysis of performance. Test equipment must also be light in weight and easily carried into restricted areas to test equipment under a wide range of weather conditions
- the product must perform the full range of tests, as specified by ETSI (European Telecommunications Standards Institute)
- a full service network must be in place
- there must be facilities to provide training of service personnel
- the operators must be convinced of the financial strength of the equipment supplier, and of their commitment to the market.

Key requirements for selling to resellers, or independent test contractors, for test of mobile units include

- low cost of product, with high throughput
- testing the main characteristics of the equipment to ETSI standards
- wide availability through distributors.

At present there are approximately 20 providers of network infrastructure in Europe, and it is usual for a network operator to standardise on a single manufacturer of equipment for test in specific areas. The new entrant thus needs to mount an aggressive sales campaign to win over network operators at the time when they are seeking to select their standard equipment. There are in excess of 2000 users of equipment for test of mobile equipment.

Announcements of product for this market have been made by three manufacturers. 'A' is the leading European manufacturer of product for GSM testing, 'B' is the leading manufacturer in the US of digital cellular test equipment to US standards, but also has a significant market share in Europe for GSM test equipment, and 'C' has the leading position for test equipment to the Japanese standards, also with a presence in Europe. All these manufacturers have a service and distribution network throughout Europe.

In addition, there are several suppliers of test equipment focused on regional markets. For example, there are two equipment suppliers whose declared strategy is to focus on equipment for the US market, and there is one Japanese company whose declared strategy is also to focus on markets for equipment to Japanese standards.

The key technology in such equipment is concerned with efficient software and hardware implementation of the various transforms necessary for analysis of the waveforms to meet ETSI standards. There has been significant progress in this area in recent years, and a significant competitive advantage can be secured by the company which can take advantage of the latest developments in this area.

The DM strategy

DM has analysed the market opportunity carefully, and has designed a strategy which it believes to be appropriate to achieving 30 % market share in this market by 2010. The company currently has a strong competitive advantage through the use of patented technology.

Key elements of the strategy will include

Maintenance of strong technology position

DM has secured patent protection, believed to be very strong, for key elements of its hardware and software for performing fast transforms. DM also has representatives on standards setting bodies, and will be aware of emerging standards with which equipment may have to comply. It has a strong internal R&D team, and works with independent consultants and academic organisations where appropriate. Considerable attention has also been paid to the human factors aspects of its designs, to make them easy to use, and DM is setting up an informal user group in order to have direct access to customer comments. An independent rating of the DM product as compared with the products announced by A, B and C, along with a summary of DM's product launch plans, is presented in the appendix.

Developing very professional selling campaigns

DM is targeting two distinct groups of users.

In the network area the company is selling large amounts of product to large users, who will standardise on its product. Although there is, currently, limited competition in the emerging market for UMTS systems, most will have product for GSM test from A, B, or C, and will have an established relationship with these suppliers. DM thus needs a very professional approach to communicating to them how it meets their key needs. It needs to reassure them about product quality and performance, and has involved them in early stage testing of product and provision of equipment for evaluation to help this process. Membership of appropriate standardisation committees and publication of leading technical papers in trade journals has also been valuable in this context.

DM has also contracted with leading independent field service companies so that 24-hour replacement of faulty equipment can be provided, and has contracted with leading training organisations in the relevant countries to provide staff training and documentation in the local languages. It has also sought to communicate the importance of the transform algorithms it uses, and has sought to ensure that tender documents issued by its customers will seek the levels of performance that it is best placed to deliver.

DM also needs to recognise the importance of the human factor in selling product of this type. The chief executive knows many of the buyers of such equipment at network operators throughout Europe, and the company is also recruiting a team of network sales specialists. The network systems buyer will also ordinarily be prepared to subscribe to an ongoing update and maintenance program, thus giving a continuing stream of income from systems sold.

In addressing the reseller and independent test workshop market, DM has adopted a different strategy. The product price point will be typically £2000, and product at this pricing level may be best suited to a direct telesales operation conducted by distributors in each country. Such distributors have been selected, backed up by a wide range of documentation and computer based training material. Advertising material has also been prepared for trade magazines. This material will have significant spill over into the network operator market, and thus it is important to ensure that it emphasises the technical and commercial strength of the company, and does not concentrate exclusively on the value for money aspects of the reseller oriented product. An experienced manager of distribution operations for professional test equipment has also agreed to join DM when funding has been committed.

Manufacturing

In the production area, DM recognises that it should outsource manufacture of product to a specialist manufacturer, and a three year initial contract has been signed with a leading UK contract manufacturer. The principal hardware areas of core technology of DM are embodied in three ASICs (application specific integrated circuits). DM holds rights to these devices, and has arranged for two different manufacturers to be able to produce them to its requirements. Testing will be undertaken by DM employees stationed at the contract manufacturer's location, and arrangements for product shipping have also been undertaken.

Company location

DM has secured office and R&D facilities near an area of considerable telecommunications industry expertise, and thus is well placed to recruit staff with specialist industry knowledge as required. Local universities also produce graduates with the rounded range of skills needed for internal sales and business development operations.

Information system

DM has designed its information system to provide real time reports of order progress, tracking of manufacturing and shipping (through links with the MRP (materials requirements planning) system at the contract manufacturer and through links with the shipping company), and reports also allow for real time monitoring of financial parameters. DM is itself a user of UMTS systems, and executives will thus be able to stay in touch with all aspects of the business and engage in videoconferencing from locations anywhere in Europe.

DM understands the importance of having strong financial backing, in order to communicate a strong position to its target customer base and to allow management to develop the business at the appropriate rate. To date, the company has secured funding for much of its development from management resources. It has also developed partnerships at an early stage with potential regional participants in this market, and has funded further development from sale of technology rights for the US and Japan to companies seeking solely to address these markets. DM now considers that it is right to take on a leading venture capital investor in order to augment its financial resources and to prepare it for a public flotation of the business in 2010.

Financial details

DM projects performance as follows, with actual results for 2004 shown for the sake of completeness. This is based on raising £2 million of equity capital now (i.e. before the end of 2005) and on raising an additional amount of £2 million in December 2007.

£million

year to end December	2004	2005	2006	2007	2008	2009	2010
sales of products and services	0	1.0	8.0	20.0	35.0	50.0	60.0
technology licence income	1.0	1.5	0.5	0.4	0.4	0.4	0.4
total sales	1.0	2.5	8.5	20.4	35.4	50.4	60.4
cost of sales	0	0.6	4.0	10.0	23.0	34.0	40.0
gross profit	1.0	1.9	4.5	0.4	12.4	16.4	20.4
selling costs	0.5	1.0	3.0	3.4	3.4	4.0	4.0
R&D costs	0.6	0.6	2.0	2.6	3.0	4.0	4.4
general and administrative	0.1	0.4	1.0	1.5	2.5	3.4	4.0
operating profit	(0.2)	(0.1)	(1.5)	2.9	3.5	5.0	8.0

Key characteristics associated with these projections are

(i)   Cost of sales is zero in the case of licence income, and is shown initially as 60 % of product and service sales reflecting lower volumes in 2006. It is shown at 50 % in 2007 and 2008, increasing significantly in subsequent years to allow for competitive pressures which may develop, and for volume sales of product which carry lower margins.

(ii) R&D expenditure is projected to grow from current levels to about £4 million annually at the end of the period in question, to allow for significant development efforts on new generation products.

(iii) Selling expenditure rises rapidly to £3 million, reflecting the need to put a sales team in place as soon as possible. It then shows modest growth as an ongoing sales commitment is maintained to build market share and to prepare for additional product introductions.

(iv) General and administrative expenditure is largely a function of operating levels.

[Detailed profit/loss, balance sheet and cash flow statements]

[CV material on promoters]

Summary

Digital Measurements has a strong technology position in the area of test equipment for the UMTS market, which is expected to show significant growth in the medium term. DM has also developed a clear market introduction strategy to allow it to enter this market and achieve at least 30 % market share in Europe by 2010.

Having read such a plan, the question is would you invest?, and what sort of deal pricing might be realistic?

Adopting the venture capitalist mindset, the main points to be considered are

Is the market interesting?

Apparently yes. UMTS looks well set for good growth, and standards incompatibility means that it is a European market. Published brokers' notes on A, B and C all seem to communicate the belief that the area of UMTS test equipment is interesting, and thus the market growth claims look broadly realistic. The report that DM has commissioned also looks plausible. Telecommunications markets seem also to ride reasonably well through recessions, which might otherwise cause worry with a company that is selling a piece of capital equipment rather than a recurring spend item.

Is the team credible?

They seem to have a good track record, and have done well to fund the business without sale of equity to date. Some gaps are present, but it seems plausible that the company will be able to get good people in the area it has chosen to locate the business.

Is the product real?

Again, DM has sold the technology rights to knowledgeable people, so this must be a plus point. However, the technology keeps moving and the report on competitive positioning of the product with other offerings may be slightly biased. If it looks interesting, the venture capitalist might, at a later stage, ask to talk with a consultant, or perhaps more relevant, with some real buyers of the products on offer. People who are existing users may be somewhat biased by the enthusiasm to justify their decision, so it is important to talk with some uncommitted potential buyers.

Can the market be entered?

The market entry challenge is perhaps the biggest difficulty. A, B and C are known to be broad range instrument suppliers, having established relationships with main distributors and network customers. DM is going to have to be innovative and have a real technology lead to crack this market. Maybe the existing suppliers are complacent or lazy, but this issue needs to be understood.

Is profitable exit possible?

If everything else stacks up, the exit should not be a problem. The margins being forecast are not out of line with those for successful instrumentation companies, and spending on product development should well prepare the company for new product launches in 2009 or 2010. The telecommunications market seems to be well placed to want new offerings, and a new wave of higher capacity UMTS systems at 155 Mbit/s may be on the market by then. If the market is looking saturated, there could be plenty of trade buyers willing to pay a good price for such a company. If there is the prospect of continued good growth, a stock exchange could be very receptive to this type of company.

The venture capitalist will then meet the management, follow up the references, check out the market with additional research, look at the financial projections and the assumptions on which they are based in considerable detail and check out key suppliers such as the chip makers and the assembly contractor. If the deal is interesting, the next step is to look at valuations and deal structuring.

6.4 Pricing and structuring

Faced with such an opportunity, the venture capital investor must look at it from the viewpoint of being able to achieve significant returns on large amounts of funds committed. The business looks as if the projections make sense, subject of course to all the commercial risks involved. If the business floats based on 2009 results it could be worth somewhere between £50 million and £100 million—say £80 million for calculation purposes—and it might be worth at least a similar amount of money to a trade buyer.

Assume arbitrarily that there will be 80 million shares in issue just before flotation, and that the price will thus be £1 per share. The business will require £2 million now and a further £2 million at the end of 2007. At that time it might be worth about £30 million before investment of the money, or £32 million after investment, i.e. 40p per share. Again counting back, this implies that there would be 75 million shares in issue after the funding of £2 million now contemplated. A price must be agreed with the founders for the transaction to go ahead.

Venture capitalists in a competitive world must sell their product. What they are providing to the founding team is money, naturally, but also a measure of support that will be of value to them through the initial difficult phases. International contacts can be cited, with the fact that pension funds of some network operators are amongst the investors in the funds, the fact that it is gaining a specialist reputation for investments in telecommunications areas and that several of the staff have direct operating experience in businesses. The venture capitalist is thus well placed to understand the issues facing the business, and is well placed to offer valuable support. Personal relationships are also very important amongst groups of people who will be quite close partners over an extended period, and a suitable venture capitalist will believe that its executives come over well in this context, as well as being seriously committed to making the venture a success.

These points of selling for the venture company are valuable in themselves, but naturally a discussion comes down sooner or later to deal pricing, and the company must also be competitive in this context. It may not have to offer the best pricing if it scores well on relationship criteria and is clearly seen as being the best on the issues of support and commitment, but pricing does have to be competitive.

Essentially, there are a number of quasi-scientific methods that give guidelines for preparing an offer to value the business, and the final valuation may be based on tough negotiations between the parties.

One guideline will be based on looking at likely rates of return. If the projections of management are accepted, or a version with some adjustments made by the venture capitalist, then some estimate of what the business might be worth after four to five years, around the time when an exit is likely to be contemplated, is possible. In this, with the proviso that everything goes according to plan, the business could be worth £80 million. Fund ranking with investors is based ultimately on rates of return, and the internal rate of return is one measure which will be applied to the fund. This measure can also be applied to looking at investments. In this case, for a venture at this stage of development, a minimum required rate of return might be between 30 % and 40 % annually. Over a four year period, a 30 % IRR translates into a multiplier of 2.9 times. A 40 % IRR over a similar period translates into a multiplier of 3.8 times. If the 40 % IRR is taken based on the projections with perhaps 7 months delay, then the shares should now be worth at most 21p. If there are 75 million shares in issue after funding, this would imply a post money valuation of the business of £15.75 million, and the £2 million investment would buy 12.7 %.

On the other hand, this might look a little rich. The implied value of the management stake at this valuation is £(15.75–2.00) million, or £13.75 million. And what is there to show for this? Digital Measurements has a well developed product, apparently, but this has not yet been accepted in the marketplace, and there are formidable competitors. Valuation can also be based on an asset calculation. The company has no net funds or other tangible assets of any consequence, having spent its resources to date on R&D and initial market probing. Trying to put a value on this is difficult. A valuation on the technology could be guessed at about £3 million, which is underpinned by the pricing level at which DM has been able to sell licences. Indeed, there is some measure of continuing licence income forecast, which might be valued (probably conservatively) at an additional £2 million. On this basis, the business prior to investment might be worth £5 million, and investing £2 million would give it a total valuation of £7 million, of which the venture capitalist's stake would correspond to a 28 % interest in the business.

And so armed, the venture capital company goes in to negotiate with the DM team and its advisors. Getting about a third of the company is known to be realistic (corresponding to 25 million shares out of 75 million, or a price of 8p per share) given the current marketplace for investment.

After some sessions, and the expectation on more than one occasion that the deal was off, eventually an outline agreement has been produced, as reflected in a term sheet or heads of agreement. The main points of this might be as follows:

Heads of agreement between Digital Measurements Limited (DM), Allied Ventures Limited (AV), J. Smith (JS) and A. Murphy (AM).

Notes: this heads of agreement is not intended to be legally binding, with the exception of point 5. It is also subject to

(i)   Full due diligence by AV on DM.

(ii) Approval by the boards of AV and DM.

(iii) Completion of formal legal documentation.

1. AV will invest £2 million in convertible preferred stock in DM, convertible into 33.33 % of the equity of DM. The stock will be entitled to cumulative accrued dividends at the rate of six per cent annually unless

converted. If AV has not sold, or been in a position to sell (based on share price criteria over a six month period), its interest in DM at a return of not less than 20 % annually, AV shall be entitled to redemption of its investment by giving thirty days notice on any date after 1 January 2011.

2. JS and AM shall enter into five year service contracts with DM. DM shall recruit a finance director acceptable to AV within 60 days of the completion of funding. Key staff insurance payable to DM will be taken out on JS and AM.

3. AV shall be permitted to nominate two directors to the board of DM, and DM shall pay fees to AV of £25 000 annually.

4. A share option scheme for senior managers over five per cent of the enlarged equity will be established.

5. AV to be given an exclusive period of 60 days during which to conduct final due diligence and finalise the transaction. DM is required not to conduct discussions with any other party during this period. All information supplied by DM to AV will be treated as confidential. AV shall not make copies of any material and shall return immediately any material to DM in the event of a transaction not proceeding.

Following the preparation of the heads of agreement, the clock starts on the sixty day period. In this period the final homework is undertaken, and work is done in parallel on drafting the legal text.

The points in the heads of agreement may need some amplification. First, it is rare to find venture investors putting in straight equity. The reason for this is that sometimes businesses fail or are sold for modest valuations—it is the real world—and having preferred stock rather than equity places the investor in a better position relative to management. In negotiations it is usually up to management to sell the positive aspects of the transaction, and thus they are weakly placed when the investor is seeking to protect his investment. If DM does poorly and is sold for £3 million after one year, then AV can get back the £2 million plus a 6 % accrued dividend, i.e. £2.12 million. If straight equity had been subscribed for 33.33 % of the business, then AV would have only got £1 million. This convertible preferred structure obviously has advantages for the investor, with management poorly placed to negotiate against it, and it has thus become the norm.

In some cases the dividend, or an interest charge, would actually be paid annually. In this case, because a good deal is believed to have been obtained on the percentage interest, AV has been relaxed on getting the running yield from this investment.

The redemption issue is logical, also. Basically, if the company has not succeeded in its main objective (and also has not gone broke in this time) at some extended date AV might want to get its money back. AV thus has rights to require repayment with accrued dividends. In practice this may be difficult, but failure to redeem just gives AV additional rights, typically those associated with a creditor of the business, or it may allow AV to take control of DM.

DM is a people business, and thus AV needs to ensure that the people in which it is investing stay with the business. Thus there are service contracts for the period in question. There is also a risk here to DM if the people involved lose interest or prove ineffective as there is a material commitment for a five year period, but AV has elected to take this risk. Insuring key human assets also makes sense, with the payment going some way to easing the disruption for DM associated with loss of one of its key managers.

Appointing directors to the board of DM is important in building the links between DM and AV. By appointing directors AV is taking responsibilities upon itself, but the act also has more than symbolic importance. The directors of a company are responsible to that company and its stakeholders as a whole. If DM gets into difficulties, the legal responsibilities of the directors from AV, just as that of other directors, must be directed towards protecting the interests of the creditors and shareholders in total, rather than having a partisan interest on behalf of AV alone. The relationship between the founders and the directors nominated by AV will hopefully be supportive, with the AV directors seeing their role as more than just monitoring and the founders seeing them as more than scorekeepers or monitors. The most valuable role for non executive directors in this context, in addition to the usual issues of corporate governance, is to assist with contacts, with recruitment, with general sharing of experience and with being industrially experienced sounding boards at the disposal of management. The fee paid is usually modest in comparison with the time commitment.

Having a share option scheme is very sensible in the case of such a company, where opportunities are to be allowed for executives other than the founders to create wealth. Five per cent might seem a little low, but there could be agreement to increase the pool at a later stage.

The confidentiality requirement is self-explanatory. The reason for an exclusive period is that if AV wants to do the deal it will have to invest a lot of time in due diligence work, and it is unrealistic for it to do this where the deal may be stolen at the last moment. The working of such a clause naturally depends on a good faith understanding being in place between the parties.

There might also be provisions about appointment of accountants and lawyers, about payment arrangements for the due diligence work, including treatment of contingencies such as the deal breaking down and the primary cause of such a failure. There might also be a requirement to appoint an independent chairman within a one year period, typically subject to the approval of the founders, with such approval not to be unreasonably withheld.

So—the due diligence proceeds. Consultants may confirm views of the product, accountants will have been through most aspects of the existing business and through the assumptions and workings in the business plan, lawyers will have checked title, references will have been taken up on promoters and a large legal document termed the share purchase agreement will have been prepared, along with service contracts, new memorandum and articles etc.

Such a document is usually very intimidating to the founder who will not have had experience of the custom crafted documents usually employed. The full document will amplify the points in the heads of agreement and will seek to cover all eventualities and seal off loopholes, thus leading to considerable length. The investor will also seek warranties from the promoters, essentially asking them to confirm that items in a long list of points about the business are all in order, except where disclosed in a disclosure letter. The purpose here is largely to flush out any points that may concern the potential investor.

So—again after heated debate on some points, the agreement is finally in place. AV and DM are now partners.

The sequels

Most stories have but one sequel. However, in this example, a number of possible outcomes can be examined. Any experienced investor will know that almost *anything* can happen after an investment has been made, and so the following sequels represent but a small selection from the range of possible outcomes.

Good news situations are dealt with first.

DM had a good start after funding. Recruitment of good people was accomplished successfully, and DM rapidly became the equipment of choice with twelve of the largest network operators in Europe, having approximately 60 % share of this market. DM did somewhat less well in the reseller type market, but managed to gain 20 % of the market after a struggle. DM raised the further £2 million at the end of 2007 at 40p per share, as expected, taking in two other investors at this stage. The company came out ahead of projections, reporting operating profits of £5 million in 2008. The real lucky break came for DM when it got the chance to acquire its US technology partner, TW. This business had been experiencing some difficulties in the production and technology application areas, but DM management knew the problems to be fixable and thus were very confident about being able to effect a turnaround. DM was able to fund the acquisition price of $10 million at the start of 2009 with debt, thus avoiding further dilution, and cash flow from the US operation and from the DM core business made such debt very manageable. DM succeeded in getting pre-tax profits from TW of $3 million in 2009, with the prospect of $5 million in 2010, and was thus able to float very successfully after results for 2009 became available, with shares valued at £1.60. AV was able to sell its interest on flotation, and reported a gain on its original investment of twenty times, with the original investment of £2 million now worth £40 million, or a gain of £38 million, with an IRR (assuming money in DM for 4.25 years) of 102 %.

Rarely do matters work out exactly as planned, but one of the sequels might be to consider a situation where this actually happened. This is still very good news for AV, with a gain on the original investment of ten times, or a profit of £18 million on a £2 million investment, with an IRR of 72 % assuming a similar period.

To give balance, some less favourable scenarios are now considered.

Just after investment, A announced an improved algorithm which effectively negated DM's technology advantage, and the technology licensees in Japan and the US sought cancellation of their technology licence agreements. Naturally, publicity surrounding these disputes spread throughout the small industry involved, and DM found great difficulty in selling product, having recruited an expensive sales force to do this. Further difficulties arose in the production area, with a dispute between the contract manufacturer and DM over availability of burn in facilities. Some dissension also arose among members of the management team. DM reported sales in the first six months of

2006 of only £200 000 compared with a budget level of £2 million, and was also evidently losing out as competitors looked close to capturing most of the main prospects among the network operators. The sales shortfall, along with having a significant selling and design overhead in place, meant that the cash position of DM started to look perilous in May and June of 2006.

An emergency cost-cutting program came too late to have any effect, and led to loss of morale. During June frantic efforts were made to secure further funding, initially from AV. AV had a major dilemma here, having undertaken a high profile investment of £2 million less than six months previously. After many sleepless nights, AV decided that it could not really commit further funds. DM had lost its technology edge, and it was now also late in seeking to sell to major network customers, which were rapidly standardising on product from competitors. Discreet approaches were made to sell DM to competitors, but these saw no attraction in taking over a situation with few assets and resources which duplicated theirs. A liquidator was appointed to DM in July 2006, with no value resulting to shareholders.

or

After DM received investment from AV, it was clear that the competitive response would be more aggressive than expected. DM had a product that was superior, but the company needed some time to convince customers that a good service network was in place, and that it was going to be around for the long haul. Also, a policy of decentralisation among most network operators now meant that local divisions had responsibility for procurement, and the approach of selling on a professional person to person basis, or on a tender basis, was no longer relevant. Local managers bought on price or on reputation, and were thus swayed by the established brand names in the business. DM's new highly paid salesforce was also uncomfortable with changing its selling approach.

After much pain, involving termination of the contracts of several members of the salesforce, DM had to make a new effort. The technological position was intact, and the company at least could count on its strengths in this area. DM had sold product to 15 of the network operators and was establishing a reasonable position in the reseller market, taking ten per cent market share in the first half year. It was, however, losing money rapidly, and the cash position by September 2006 was very tenuous. AV decided to invest a further £2 million at a similar price of 8p per share, taking its interest now to 50

%, although effectively still being in a minority due to existence of the share option scheme.

The story improved in 2007 with smaller losses, but still DM was evidently struggling. An approach was made by B which had respect for DM's technology and which needed to strengthen its presence in Europe. DM was sold to B in early 2008 for £9 million, which gave AV a modest profit of just over £500 000, considering the cumulative dividend on the preferred stock.

Making venture capital investments, and indeed being the recipient of venture capital, is an interesting business, as the examples illustrate!

6.5 Stock exchanges

Stock exchanges will be perceived by many as being dominated by large companies, with the bulk of turnover coming from trading shares rather than from companies raising money. There is, however, often investor interest in smaller company shares where there is the realisation that smaller companies generally have greater upside, in that they can often be well placed to double their revenues within their market area. For this to happen in the case of a larger company the overall market would have to grow by a considerably larger figure. Smaller, focused companies are often seen as being at the start of some new emerging area of significant interest, with the prospect for high returns. The expert investor with an understanding of the technology and market issues involved can put specialised knowledge to good use in studying smaller listed companies, and investing accordingly. Investor appetite for small companies, often in 'technology' areas, can allow them to raise money on stock exchanges at what may appear to be very satisfactory valuations. A particular attraction for investors in companies with a stock market quotation is liquidity—the ability to buy and sell shares at any time, at a price determined by supply and demand issues. In practice liquidity may be somewhat limited in the case of smaller companies, but some measure of liquidity is often regarded as better than being locked-in to an investment in a private company.

All investors will be fickle. Venture capitalists will regard some sectors as promising at any given time, and likewise only more so in the case of the stock market. A herd instinct may be present on the part of many investors, with everybody wanting to enter some area such as communications technology or the Internet one day, and then the next day regarding the whole sector with some mixture of caution and derision because one company in the group has come out with poor results.

Investors are probably particularly fickle about technology stocks, and thus shares in such companies are volatile, often with wild price swings based on rumours or product announcements. Information about companies has to be given by public announcement, in order to avoid creating situations of unfair trading opportunity, and thus there is no time for reflection on the part of investors before trading can lead to dramatic rises or falls in stock prices. Good news or bad news can have disproportionate effects on the share price; it all comes down to investor psychology. Every company has its strengths and weaknesses, and it is possible in all cases to look at the glass as being either half-full or half-empty. Good news prompts all the upside associated with the company to be viewed, while bad news rekindles all the worries that potential investors might have had. Such companies usually conserve their cash by payment of no (or very modest) dividends and this naturally increases volatility.

It has been conventional for companies, certainly in the UK, to wait until a significant record of profitability had been achieved before attempting flotation, but research stage companies, such as Magnum Power in 1994 or several biotechnology companies during 1992 and 1993, are occasionally floated based on having an experienced team with a technology position which appears well protected. The appetite of public market investors for companies in such a position may well increase, given renewed interest in smaller stock markets across Europe (e.g. the AIM in London and Le Nouveau Marché) and with the pending launch of EASDAQ, a pan-European market likely to be particularly suited to technology companies. A key requirement for success of market segments populated with early stage technology companies is a good resource of stockbroker analysts, who are able to give a balanced view of companies to investors and encourage participation by institutional investors as well as individuals.

Readers who wish to pursue the matter further would do well to subscribe to some of the publications or tipsheets focusing on this area (e.g. *Technivest*) which will give numerous examples of the real issues affecting quoted technology companies.

In summary, a company may be advised to seek a quotation if

● it can cope with the reporting requirements and disclosures associated with being publicly traded, and the very real costs in financial terms and in management time

- it expects to benefit from the extra standing associated with a public quotation
- it is prepared to live with situations where the share price may be very low on bad news, which may be just a temporary hiccup
- it can identify a broker with good analyst coverage who will be able to present the shares well to long term oriented investors.
- there is either a good recent trading record, ideally with profits exceeding £1 million annually and with expectations for good growth, or it can convince brokers and later investors that it is very well placed in terms of technological strength and/or management capability.

There is a very wide range of approaches to valuing quoted businesses by investors, particularly those perceived as being at the early stages of becoming significant businesses.

P/E—or price/earnings—ratios are usually given on a daily basis for most quoted companies in the financial pages. The P/E ratio is usually the only quoted statistic, and it thus acquires a lot of importance in the eyes of many investors. It is simply the ratio of the current share price to historical earnings per share. A prospective P/E might be quoted based on the ratio of share price to expected earnings for the company in the year in question.

Using a single figure such as P/E ratio in seeking to analyse a company is naturally a rather crude approach. Earnings may be depressed by high R&D costs, and companies may elect to deliberately operate near breakeven or make losses for extended periods in order to increase their market share. This, of course, always assuming that they have the cash resources (or access to cash resources) to fund development without having internally generated funds. P/E ratios are thus most meaningful in the case of companies which are operating with a more stable allocation of resources to R&D, market development and other promotional activities.

In general, however, the market tends to assign P/E ratios to stocks based on the type of earnings growth it expects from them. At the bottom, perhaps on P/E ratios of four to six, are situations such as car companies at the top of an acknowledged industry cycle, where industry analysts see earnings growth as being negative as soon as the cycle peak has passed.

P/E ratios of between ten and 20 might be applied to most companies experiencing steady growth, with figures at the top of this range being assigned to companies operating in areas perceived to be growing much faster than the economy as a whole. Many technology companies, in good times, can aspire to ratings in the high teens or indeed much higher depending on the hope value. Companies with strong balance sheets might also be rated somewhat higher than those obviously creaking, but on the other hand companies using debt efficiently to give good return on equity characteristics will be praised by the market.

A more difficult issue arises with concept companies, or companies which seem well placed in a market that is going to be big, but which for the moment is quite modest. This issue has been encountered in 1995 with the Internet boom, as Netscape Communications, UUNet, Spyglass and other companies were allocated market capitalisation figures totally out of keeping with current performance.

In such cases, analysts take a long term view of the market. The Internet market, worth about $1 billion in 1995, is widely seen as worth about $20 billion in 2000, and to be still growing quickly at that time. If a company is well placed to be the market leader in five years time in a fast growing $20 billion market, then this is obviously reflected in current valuations.

If, for example, a company is expected to be capable of making $500 million in five years' time, but is now breaking even, how is it valued in the stock market? One approach is to suggest that if it is making $500 million then, it will then have a P/E of at least 20, as it will still be growing, and thus will be worth $10 billion. Looking forward, it will be recognised that things don't always turn out as planned, and thus expected valuation might be discounted by 30 % annually. Working back five years, this might give a valuation of $2.7 billion today, even though the company now might only be breaking even on small revenues. The suggestion is not that the reader embraces such logic with great enthusiasm—but it is important to realise the valuation mechanisms that can be used by those looking at companies.

Other valuation mechanisms relate to activity statistics. Cellular telephone operators, alarm station monitoring service providers and cable TV companies have typically been valued on a price per subscriber basis. This reflects the acquisition pricing paid by industry consolidators, who would naturally look at the alternative options of buying a company with a block of subscribers or putting in their own infrastructure and persuading or incentivising customers to sign up.

Companies may also be valued based on assets. These assets can include real property, in the case of property companies, or have an element in the valuation for intellectual property, as may apply in the case of a company with significant patent holdings and/or valuable brands. In such cases, value may be recognised by the market only in the case of a contested takeover bid.

Section IV
Developing the business

Chapter 7
Alliances

Alliances come in several forms, and include arrangements such as distributorship agreements, technology sharing and licence arrangements, joint ventures and shared buying groups. The objectives of parties in entering into an alliance are furtherance of their individual objectives, and alliances may continue to exist only for as long as they continue to bring benefit. Alliances may be tactical in nature, as when two companies come together to form a consortium to bid for a large project, or they may be strategic, designed to be in place for a longer, perhaps indeterminate, timescale. Small companies can be motivated to enter into alliances with larger companies to exploit the distribution channels offered by the larger partner, and larger companies may seek to draw smaller companies into alliances to gain access to unique technology or other strengths.

Alliances may be formed for market development purposes or for defensive purposes. The company repelling an attack from a newcomer will seek to control potential partners in a wide range of areas, from suppliers to downstream distribution channels, to other potential partners offering complementary services. Understanding the true motivation behind those seeking alliances is thus important.

7.1 Types of alliances

Alliances can be classified as follows

7.1.1 Distributorship agreements

A low cost way for a firm to enter into an unserved geographical area may be through appointment of a distributor. The distributor is chosen on the basis of being able to add value to the product, which can include providing better logistics, technical support and customisation, as well as playing an active promotional role. In some cases the distributor may be willing to buy rights to a territory for money up front, which is of particular appeal to the cash limited early stage firm. The manufacturer is trading the incremental revenues and margin derived from such activities for constraints and limitations as follows

(i)   Control over customers and contact with customers is ceded to the distributor. The distributor has gained power through this customer contact, which may be transferred to benefit a competitor if the loyalties change. The manufacturer is also somewhat insulated from customer feedback.

(ii)   The distributor is educated in the industry. This implies the risk of subsequent entry of the distributing firm into the industry as a competitor, and one who knows its opponent's weaknesses.

(iii) The economies of the business can shift easily, and a territory that was unimportant may now have increased in importance, or the company may now be able to afford to cover territories directly. In this case the company will probably want to go direct, but may find its freedom constrained by distribution agreements. The initial distribution agreement had to protect investment by the distributor in developing the market, so usually there is some notice period required for termination.

Distribution agreements are by their nature specific and complex. They include consideration of new products, margin sharing and other protections for manufacturer and distributor. Designing distribution relationships must be done with long term issues in mind, otherwise the future may be unduly constrained.

7.1.2 Technology sharing and license agreements

It is often beyond the resources of a company to undertake the R&D work associated with a project by itself. Technological capability can sometimes be included in items bought in as industry standard products. The personal computer manufacturer, for example, buys in industry standard integrated circuits, a disk drive, monitor and power supply, and uses an industry standard operating system. In many cases a ready supply of complementary industry-standard products is not available, and the company has to team up with another company for provision of specialist high technology items. Formal technology sharing or collaborative research may be essential if the company is to be a serious player in the industry. Projects such as development of digital cellular radio, development of new drugs or developments in biotechnology may be of a scale beyond that which can be comfortably handled by most companies within an industry, and thus the push to collaborative activity.

Licensing process or product technology can be attractive, as it allows a firm to gain revenues in areas where it might otherwise have been unable to operate. In industries where it is important to set standards and/or where additional sources of product are demanded by most customers, then licensing cheaply can promote the approach. In a situation where the company has a proprietary technology that can be incorporated cheaply by many end users at a price that does not justify their developing their own approaches, it can have a very attractive out-license business, as exemplified by the Dolby Sound noise reduction system.

Often competitors are needed to help develop the business. If one company is the sole source provider of product, then it may not be taken seriously. Thus the motivation in the semiconductor industry, less so perhaps now than in earlier days, for ensuring that any newly launched device had a promised second source, in order to give designers some extra comfort in designing in the device. The semiconductor industry has succeeded to some extent in recent years in getting designers, and corporate managers who develop such policies, to accept singly sourced products, pointing out that the delay in waiting for products to have multiple sources implies a competitive disadvantage.

In situations where the approach is licensed out to someone who will use it to enter the market, the company needs to be cautious. It is educating a potential competitor. It may be that the emergence of this licensee as a competitor is inevitable, in which case it is better to provide the tools for its attack than have another competitor do so.

A licensor expects from a license agreement

- incremental high margin royalty or lump sum income
- increased prestige for the technology, and greater acceptance as the industry standard
- being able to offer an extra source for its products to reassure buyers for whom continuity of supply is essential
- increased market access, particularly into countries where giving a license is a prerequisite for operations
- being able to profit from innovations made by the licensee
- faster movement down the experience curve as the cumulative volume of products involving the technology grows more quickly than with a sole operation.

The risks to guard against are

- misuse of the licensed technology
- the commitment to license technology on an ongoing basis, with perhaps limited power as the licensee can threaten to switch to competitors.

With these situations, some paranoia is appropriate, as competitors can emerge quickly. The big risk of licensing may come about through introducing a competitor to an industry, rather than from abuse of the technology. Given the change in many technology cycles, the next generation which the licensee may develop may be well independent of what has been licensed.

There is also the situation where alliances change, with licensed technology likely to go directly to a competitor as new joint ventures are formed or as a licensee is acquired. This has been a fear in the aerospace industry, for example.

7.1.3 Supplier groupings

It is common for strategic alliances to develop between a leading firm and its suppliers. This has been seen in areas such as the Japanese car industry, where a network of suppliers is virtually dedicated to a single customer. The dependence is acceptable to the suppliers because they receive assurances of continuity and benefit from technology and skills transfer, as well as having the simplicity advantages of dealing with one customer.

Similar supplier groupings are being developed by major retailers, for example in the clothing area. Indeed, there is a broad trend to vendor base reduction, which automatically brings about a mutual dependency between firms, and this translates effectively into a strategic alliance. The former logic where companies were reluctant to have suppliers dependent on them for more than ten per cent of business has effectively been turned on its head. This approach was consistent with an adversarial relationship having short term characteristics, rather than the longer term win-win and shared savings environment that is now being developed with suppliers.

7.1.4 Joint ventures

Joint ventures are probably the ultimate form of cooperation, prior to total merger. The motivation for a joint venture comes from issues such as

- compulsion, as in many countries a joint venture with local interests is required
- combining of corporate skills. An example was the Prodigy service operating in the US which is a joint venture between IBM and Sears in providing a home user oriented electronic shopping, mail and travel service. This network requires the computational skills, presumably brought by IBM, and the retailing skills and buyer understanding presumably brought by Sears
- as an aspect of tapered integration. Rather than integrating backwards fully and taking over the role of our suppliers, it is useful to be able to form a joint venture operation which assures supply to the company and which uses industry knowledge of the supplier. This arrangement may lack stability, if for some reason the company can buy more cheaply elsewhere, or if the supplier feels that other customer contacts are being impaired by being allied with one customer.

A joint venture has its own persona, typically with executives seconded from the sponsoring companies, and it may accumulate significant fixed assets and market position. It is thus set up from the start to be a somewhat more permanent entity than other forms of strategic alliance, requiring more care in its implementation.

7.1.5 Franchising

Franchising has grown in popularity, particularly in the retailing area. It gives the franchisor

- more capable and driven management at the local level than is provided by employees. In businesses where a local identity is important, franchising may provide a means of ensuring that this is provided
- royalty income in territories which it might not otherwise serve
- in industry race situations, rapid expansion with lower capital requirements
- in addition to the franchise fee income, profit from sales of product.

As with all alliances, the terms of the franchising agreement are central to whether the arrangement will work. The franchisee is making an investment in hard work and in capital in funding the operation, and needs to have some assurance that this effort is protected against competition or withdrawal of support by the parent.

Franchising is particularly appropriate in areas such as retailing, where the benefits of a big organisation can be combined with the hard work of the owner operator. The big organisation can bring brand recognition and a perception of high, or at least uniform, standards in the mind of the customer, as well as some economies of purchasing. This approach is well suited to restaurants, hotels and similar service businesses.

7.2 Focus

There is an increasing recognition on the part of larger companies that undertaking all activity in house is rarely appropriate. IBM has, for example, been very active in forming alliances with Apple, Siemens, Stratus and other parties for joint marketing and/or technology development. The sheer cost of large technology development projects such as GSM (digital cellular telephone) or the new generations of DRAM memory devices make co-operation quite essential.

On the part of large corporations there is also an increasing move to outsource activities other than those associated with areas of core competence. Core competence is most likely to be found in product design, in some specialised manufacturing areas with unique process technology and in customer contact and marketing areas. Out-sourcing all other activities makes sense. It is a strategic decision, involving considerable mutual dependence between the partners involved, as it is very difficult to reverse outsourcing decisions or to switch rapidly to other outsourcing partners. When done properly, outsourcing is also bound to have many of the characteristics of an alliance.

A new attitude towards co-operation is also consistent with the approach of large companies of breaking up internal monopoly divisions and exposing them to competition. Such an approach leads to a more cost based attitude to justification of projects. Coupled with a reduction of the self confidence (or perhaps arrogance!) of larger companies, co-operation makes increasing sense.

Alliances can be viewed as being along a vertical axis, looking at issues of quasi-integration forwards or backwards. Alliances with distributors and franchisees can be viewed as getting some measure of control downstream from current activity, and those with suppliers, whether of goods, technology or other resources, can be regarded as going upstream to secure control of key dependencies in this area.

Allying horizontally gains a broader range of capabilities. For making instruments, it is desirable to be allied with suppliers of complementary products to offer a complete solution. If producing metalwork for the computer industry, it may make sense to ally with a plastics manufacturer to offer a full range to customers.

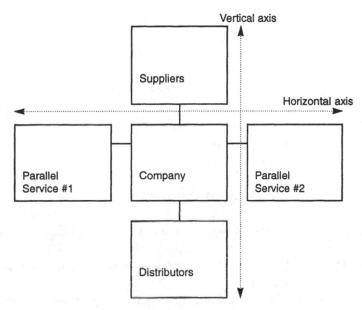

Figure 7.1   Upstream, downstream and horizontal alliance opportunities

Current thinking in relation to strategic procurement (e.g VENKATESAN, 1992; SPEKMAN, *et al.*, 1994) regards suppliers as effectively in alliance with the customer.

The duration of alliances

It is important to realise that business conditions change, and the structure of alliances will change over time. Some alliances may be short term marriages of convenience (acknowledged as such *ab initio* by all parties) and others may be enduring relationships between businesses which last for decades. The important thing is that alliances are reviewed openly at periodic intervals and are not regarded as immutable. Alliances will come under strain for numerous reasons. A company may want to control its own sources of supply or its own distribution. It may start competing in with the alliance partner in some peripheral areas, which gradually assume more relevance. An acquisition by the company or its partner may cause strains in the alliance.

In practice, alliances can be expected to be of modest duration in turbulent industries. Alliances in the computer industry, in particular, have shown this characteristic in recent years. Alliances between otherwise serious rivals can be possible where the common enemy is seen as a greater threat. Thus there have been alliances aimed at attacking the strong positions held by Intel and Microsoft in PC hardware and in PC operating systems respectively.

The average age of alliances is about seven years (DRI EUROPE, 1994), with some lasting much longer than this. Examples of successful long term alliances include the Ford-Mazda alliance, and the Fuji-Xerox alliance in copiers.

7.3 Equity underpinning of alliances

In some cases, participants may seek to cement an alliance by having an equity investment by one company in another. This can create initially a feeling of sharing of future success, but is not in all cases the most appropriate course of action.

The first point is that the commercial fundamentals in themselves must be of mutual attraction to the parties involved. If this situation does not occur, then having a shareholding will not change the picture and can act as an impediment to restructuring the alliance for the future. If the alliance holds together because of an equity bond after the commercial logic has evaporated, it can be a distraction to both parties. Realistically, however, all alliances go through their periods of questioning. When the initial euphoria has subsided, there is frequently a need for both sides to work at making the alliance work. The will to make it work will be influenced by investment already made in the alliance and by the future opportunity. As equity participation is likely to have increased the amounts involved here, at least for the investing party, then there is a greater incentive at least for that company to want to make it work effectively.

Associated with significant equity participation is also the issue of board representation. This implies a sharing of secrets and plans with the alliance partner, and may make change very difficult, if circumstances change. The existence of a significant minority shareholding can make the company significantly less valuable, particularly if the shareholding has any special rights.

Apart from small amounts by way of passive investment, equity investment of this type has to be regarded as a fundamental change of control, possibly the first stage of an acquisition of one company by another. Such an investment has therefore to be discussed rather seriously by the company selling the minority stake. If the alliance can be operated commercially without the complications associated with this investment/control issue, then it makes little sense to incur them.

7.4 Alliances and the market life cycle

In DRI EUROPE, 1994, a table is given which outlines the main criteria for competitive advantage at various stages in the market life cycle, and the corresponding goals of strategic alliances.

These are cited as follows

Market life cycle	Competitive advantage—main criteria	Alliances goals
Development phase	importance of technological innovation	access to new technologies access to market
Growth	market access to finance innovation	access new markets build market position differentiate to survive
Shakeout	technological process innovation	build market position differentiate to survive re-establish critical mass
Maturity	importance of cost reduction	re-establish critical mass achieve cost leadership
Decline	withdraw	withdraw

7.5 Selecting alliance partners

The first point here is to establish the rationale for an alliance. Once the company knows exactly what it is seeking, then the search is for the partner who will offer best fit, allowing the entity to achieve its objectives and without excessive compromise on issues such as control and future development.

An area in which strategic alliances are becoming important is the telecommunications service industry. This industry was characterised up to the early 1990s by state run monopolies in most countries, with the US and UK leading introducion of competition into the industry.

In the mid-1990s, a number of large groupings are forming. These include the Global One alliance of Sprint with Deutsche Telecom and France Telecom, the BT/MCI alliance termed 'Concert', and the Unisource alliance of Dutch, Swedish, Swiss and Spanish operators, etc. The effective consolidation of the industry into groupings of dominant or monopoly national operators leads to charges of cartelisation. In this environment, the national operators in the smaller countries are looking at these entities with a view to choosing partners.

In all these cases, the principal points motivating the strategic alliances are

- gaining strength in a consolidating market
- being able to offer international services to businesses.

Selecting partners based on fit is very desirable. The greater the commonality in objectives, in culture, in organisation structure, in use of information systems, the easier it will be to work together and the less the risk of misunderstanding. The fit may also come from a recognition of shared need. If both organisations really are likely to gain from the alliance, if they are complementary in most areas rather than having large areas of overlap and many conflicts of interest, then there is a greater chance of making the arrangement work. As with all alliances or coalitions, there are trying times, and greater fit is likely to see the alliance survive.

7.6 Making alliances work

Alliances work at several levels within an organisation, and are not just based on a piece of paper having signatures of the chief executives of the respective partners. Alliances need to work through an organisation. If the headquarters of a geographically dispersed business says that a service provider is to be used from now on in each location, then this may cause hostility. In a situation where people are being empowered within an organisation, it does not make sense to foist an alliance upon them. The rationale for the alliance has to be made clear, particularly if it involves changing suppliers at local level. In a small community there may be strong links with the existing suppliers of service and the logic for incurring the pain associated with a change of supplier must be communicated well. Or, perhaps, such factors mean that the alliance was not appropriate in the first instance. Some of the issues involved are treated well in KANTER, 1994 and in LORANGE, ROOS and BRØNN, 1993.

Several examples of alliances that are finding the going difficult can be encountered in the computer industry. This is a fast moving industry, and there is a perpetual need for companies to back the winner in terms of setting standards. There were alliances between IBM and Microsoft in the 1980s concerning DOS and early versions of OS/2. IBM and Apple have been allied in various projects in the 1990s. The PowerPC chip was successfully developed, but the follow through in terms of a common operating system and other support software has been somewhat delayed. The software joint ventures, Kaleida and Taligent, were also dogged by issues concerning specification of objectives, and by competition with other units within the partner companies.

The fact that some alliances do not succeed is perhaps unsurprising and indeed may be no bad thing; in a changing environment companies will want to keep open a number of development paths, just as they would keep open a number of development projects. It is probably easier to fold an in-house development activity that finds itself on the wrong road as compared with ending an alliance, and it evokes less publicity!

7.7 Competition between alliance groupings

In several areas of industry, competition can be found between groups or camps of companies. This phenomenon has been well described in GOMES-CASSERES, 1994. A particular aspect of this group competition can be based on standards, with each group effectively pushing a rival standard. The RISC microprocessor has been identified as an area where alliance networks have dominated the business environment, with various alliance constellations centred around MIPS, Hewlett Packard, IBM and Sun. Joining alliance networks can have a defensive aspect, or it can be simply a case of jumping on the bandwagon. On the other hand, alliances, particularly if the wrong bandwagon has been chosen, can limit flexibility.

7.8 The unintended sale as a consequence of a strategic alliance

A strategic alliance represents a close coming together of two companies. Alliances are probably best placed to work well if the two companies bring together complementary strengths, and if these relative strengths look set to continue for an extended period.

In other cases, there is usually a clearly identifiable weaker party and a clearly identifiable stronger party. The weaker party can often get weaker, becoming more dependent on the stronger party. Particularly if the stronger party has an equity interest in the weaker party, the situation may arise where the weaker party is not really saleable on the open market, and can be acquired for much less than what might have been full market value by the stronger partner (BLEEKE and ERNST, 1995). As a counterexample to this, one can cite the Rover Group, which derived significant benefit from the Honda association, and which was eventually sold to BMW, but this may have been as a result of very good management of the alliance by Rover or by Rover's owners, BAe. In cases where the alliance is managed less well, drift into a weaker position can be unavoidable.

7.9 Managing alliances

Alliances can give major benefits to a company. They can expand its geographical reach and coverage, and they can allow suppliers to ride on the growth paths of their customers. In the case of technology companies, intellectual-property alliances may mean greater resources being available to protect this property and to establish it as a standard.

Alliances bring with them an opportunity cost. If an alliance is to be meaningful, it is usually with one partner in a certain area, or at most with a small number of partners. If the partner has been chosen wrongly, then the growth can be limited, and the company has the double difficulty of seeking to extricate itself from an unsatisfactory situation. This may involve a direct cost in terms of compensation liabilities, but perhaps more costly may be the disruption in business during the transition.

Thus alliances need to be managed. They are too important—in terms of upside and in terms of subsequently tying the hands of the company—for drift to be permitted.

7.10 Alliances versus acquisitions

The ultimate alliance could be said to be a merger, or the acquisition of one company by another. There is evidence, however, that alliances can work rather better than acquisitions. With alliances, there is still a strong measure of independent management within the units choosing to ally, whereas a completely different management approach may be required if the businesses are to merge. There are the corporate culture issues here. It is possible to accommodate the differences in corporate culture necessarily associated with alliance partners as long as there is only an alliance, whereas an acquisition can less readily accommodate such differences. With an alliance, limits in the areas of collaboration can be specified. These may be in clearly defined areas, for example in pre-competitive research with a view to setting future standards, or serving one particular customer who has forced the alliance in the first place. Outside the realm of the alliance the companies can compete freely as before. In an acquisition, most managers cannot harbour the idea of competition between divisions, and the situation of competition coexisting with collaboration may just not be possible.

Acquisitions can also be very expensive, and availability of capital to undertake acquisitions may not be present. In this case, a network of alliances can be a substitute for a fully wholly-owned network of businesses. The key factor here is to get the alliance members to present as seamless an interface as possible when they deal with potential customers.

Alliances have been a very topical issue in management journals in the 1990s. Relevant papers include those by TURPIN, 1993; SASAKU, 1993; PEREDIS, 1993; GARNSEY and WILKINSON, 1994.

Chapter 8
Acquisitions

8.1 Motivation for acquisitions, and key issues involved

Acquisitions can be a key element in corporate development. A catalogue of acquisition types outlines the issues involved.

8.1.1 Financial acquisitions

These are acquisitions done solely for financial considerations and not as part of an overall strategy, which would be partly reliant on synergy between the acquired business and current operations. These acquisitions may be made for earnings growth or to achieve some measure of financial diversification of an overall portfolio of businesses. A company on a high rating can secure earnings per share growth by acquiring businesses on a lower rating, and paying for these by issuing paper or raising cash based on its high rating. This approach has been a key element in building conglomerates, aided until recently by accounting rules which allowed significant provisions to be amassed in the case of acquired companies.

Acquisitions of this type are also undertaken by buyout funds. Frequently there is a focus on management involvement, in which case transactions may be termed management buyouts or management buy-ins, and these topics are treated in more detail in Chapter 9. There are more general buyout funds that spot an opportunity to take over a company, perceived to offer an opportunity for creation of value above the required purchase price. The value creation may come from brand building, from cost cutting, from repositioning, or any of a number of techniques which can be used to make the business appear more valuable at the end of a holding period of a number of years. Examples of successful acquisitions of this type include the RJR-Nabisco transaction, the Snapple beverage transaction and others, with ANSLINGER, 1996, treating the issues involved more completely.

8.1.2 Strategic acquisitions

Strategic acquisitions are those which are driven by a clear strategy, usually implying that some synergy is being sought between the business to be acquired and other businesses currently owned, or likely to be owned, by the firm.

Strategic acquisitions come in numerous types, undertaken for different purposes. These are summarised as follows.

Gaining market share

It is often logical for two competitors to come together to gain a stronger market position. This type of acquisition is naturally examined most closely by competition authorities, to determine if the acquisition will be considered as unduly limiting competition in the marketplace. As industries consolidate from a plethora of firms to a small number, numerous acquisitions of this type will take place.

Repositioning

A company may decide to reposition its businesses; the typical motivation here is to achieve some focus in the overall operation. A company might decide to sell off businesses in which its strategic positioning was weak or which were going to show low growth, and try to grow in areas in which a strong position was possible and which were seen as areas of high growth opportunity. The implementation of this approach usually requires acquisitions to strengthen position in the areas viewed as attractive. An example of repositioning of this type is the Reckitt purchase of L&F in the household products area in the US, along with a sale of the formerly core Colman's mustard business.

Operational diversification

This is where a company may wish to get a better spread of business. A printer serving the computer industry might find it desirable to buy a printer serving the automobile industry and combine operations. Serving different customers gives some measure of operating diversification. In some cases a company will seek to get a business which is directly counter seasonal to its own. A school book printer is very busy during the summer and might find it attractive to buy a business specialising in catalogues which could be produced in the winter months.

Integration (forward or backward)

It can be tempting to acquire distributors, or to acquire suppliers of a key component. These businesses are usually partly known quantities as the company will have been dealing with them for some time. The difficulty here is that a business is often competing with its customers or suppliers. Sometimes the threat of being able to integrate the business up or down the chain can prove attractive in getting better terms from suppliers or from distributors of products. Sometimes a participant finds it necessary to take decisive moves to change the integration pattern of an industry. In the computer industry, personal computer suppliers periodically alter strategies by going direct or later declaring a policy of selling only via independent dealers. The debates in 1994 between Intel and major PC manufacturers such as Compaq illustrate the issues involved here.

Resource acquisition

Frequently the motivation behind acquisition of a business lies in getting control of some asset of that business. This might be its customer base, the employment arrangements with key staff, the technology or brands involved, or indeed physical assets such as plant, timber resources, minerals or oil, or land. When acquiring a small software consulting company business, for example, it is the right to effectively recruit the people onto a payroll that is being bought. For buying a drug company, the focus is usually on the product portfolio and on the pipeline of work underway and at various stages in trials. A company buying a branded goods manufacturer is likely to have to pay a premium for the brand value, a factor recognised in current efforts to value brands. Taking over a medical practice or a cellular telephone business is really a way of taking over the customer list and thus supplying services to that customer group in the future. Likewise, a business might be acquired for the plant and machinery it has or maybe for some land that can be put to better use.

Geographical

This particularly applies to distribution intensive businesses and to other local service businesses. Distribution businesses find it particularly difficult to establish greenfield operations, and there is often very little that differentiates one distribution business from another. In mature markets customer loyalty may be a strong factor, and the established player may well have a dominant position in a particular area. To open up a greenfield operation invites several years of competitive attrition, with the newcomer starting from a zero base, and with no guarantee of success. Thus there is a strong motivation for acquiring businesses if a company seeks good geographical coverage.

Bolt on or infill

These acquisitions are probably the simplest to make. When consolidating a territory they represent the final phase after the major acquisition has been made. They may represent the acquisition of a single store by a growing group. Usually considerable synergies can be obtained.

Window or beach head

The only way to learn about a market is sometimes to undertake an operation there. Frequently companies undertake a small acquisition in a new area to sample it prior to committing major resources. The important thing of course here is to be objective at the end of the trial period as the company may have particular champions of the experiment in question. Frequently a beach head in which a company has a small market share position is not a real test of what could be done with serious participation in the industry. Acquisitions of this type are sometimes found in geographic markets that are new to a company. They may also be undertaken in areas that are 'faddish' at any given time.

Exposure

Similar to a beach head operation, this has the objective of providing a learning experience for the entire operation. It is often very useful to gain exposure to a new and demanding market if a company wishes to strengthen its operations. This is the approach cited by Michael Porter (PORTER, 1990) as being used successfully by companies seeking to develop as significant international participants. A company may often be dominant in its home country. If it is to become a significant international participant, it makes sense for it to enter the most competitive, demanding market for its products or services, and learn as much as possible. Acquiring a company in such a market, with a view to learning from it, may be an important element of strategy.

8.2 The value of an acquisition

Pricing of an acquisition transaction is dependent on a number of factors. The acquisition is naturally expected to be compatible with the longer term objectives of the acquirer, and so must be judged on longer term financial criteria.

The pricing of a financial acquisition, i.e. one done without great expectations of synergy with the rest of the business, will probably be based on other investment opportunities. The return on a similar amount of money invested in public stock markets must be the benchmark, and price is based on deal statistics which are unlikely to depart much from those of comparable companies having a broadly similar growth outlook. The acquiring company will not want to pay much more than public market multiples as it is not looking either for synergy or the need for control usually associated with getting synergy.

The pricing of the other types of acquisitions is dependent on the assumption that the buyer is able to gain by integrating the new business into his own. The buyer may pay any amount that is consistent with improving long term position.

There is the belief that long term profitability is correlated with market share, as substantiated by the PIMS studies (BUZZELL and GALE, 1987). Two participants may be chasing market share aggressively in an emerging market when a significant independent company comes up for sale. Whichever company acquires this independent firm then gets a commanding lead in the race to consolidate the market share position in the industry. The independent company can profit from being in this position and can effectively be auctioned between the two bidders, both of which are willing to take a long term view of the likely valuation of their resultant market share. The valuation is thus likely to far exceed any figure that is indicated by conventional financial ratios. Areas of the communications industry, such as cellular telephony, have seen acquisitions made based on this sort of rationale.

A similar type of bidding war can take place for technology and for brands. In these situations there is as much a desire to deny the purchase opportunity to a rival as to take on the business. The bidding war between Sainsbury and Tesco for William Low indicated the intense interest in securing their respective positions in Scotland. The battle for Spear's games, owner of Scrabble, reflected interest in a unique product. Frequently in races of these types, to lose may be seen as positioning the company as an also ran. This is usually a particularly unattractive position from the management perspective, and may fuel the acceptability of a higher price if the prize is to be secured.

Sometimes the valuation calculation can be quite straightforward. Allowing for synergy with existing operations and expectations of better management under new ownership, the profitability of the operation in these circumstances can be calculated. A multiple of earnings appropriate for the industry is applied, giving the value to the new owner. The costs of implementing change can be subtracted and the resulting figure is what the prospective new owner can theoretically afford to pay.

This may reflect a very significant premium on any valuation of the business suggested by normal criteria based on financial performance, and may be a significant premium on the stock exchange value of the company being acquired, if the market has not factored in the possible strategic values of the business in question. This means of course that the vendor benefits from improvements that the new owner may expect to be able to bring about, but in a competitive bidding environment the bidders may be prepared to pay the price.

Likewise, when buying a beach head, a company may be prepared to pay quite highly as the purchase price may be not material in the context of its overall plans for the sector or territory. On the other hand, when undertaking a beach head acquisition, the target is unlikely to be unique and the purchaser may be able to shop around, particularly if it is looking for one of a multitude of companies in some broad geographical region. This applies less so in the case of the exposure acquisition. Here a very good company is being sought, usually in a specialised field, from which to learn, and opportunities to shop around may thus be more limited.

Acquisitions are also very influenced by timing issues. If value can be seen at a time when disappointing reported performance from a company has soured the market, there can be a significant value building opportunity. Likewise, in the case of a business owned by an entrepreneur or by family interests, there will be occasions when the owners are more predisposed to hear acquisition proposals than at other times. There can be times when an investor in the business needs liquidity. Many of the most successful acquirers stay in touch with a small number of acquisition targets over an extended period, seeking to be the first port of call if the owners do decide to sell.

This is, of course, the picture as viewed from the buyer's side. From the seller's viewpoint, there are other issues to be considered. Not to be dismissed is the idea of finding a good home for the business. The entrepreneur or family business owner may take the view that sale is appropriate for some reason. The industry may be consolidating, the imperative may be one of get bigger or get out, and resources may be such as to favour being bought rather than embarking on an acquisition program. In this case, the owner is more likely to sell to a company which will maintain the business in the locality, will treat employees favourably and will let him retain respect in the local community. If the buyer can profit from this good home desire, so much the better. This perspective also applied to many of the sell-off situations in Eastern Europe, where the selling government was keen to find buyers who would invest in the businesses involved and maintain as much sustainable employment as possible.

There may be situations of sale being effectively forced on a company. The consolidation process may make the acknowledged future for the smaller firms in an industry quite precarious, and the seller may be well aware of this. In this case the seller may not be in much position to engineer an auction position for the business. In situations where a large number of companies of the same type are sold in a business area without much opportunity for differentiation, then rules of thumb for deal pricing get established from which it may be difficult for the buyer to get any major deviation. An example might be 'five times EBDITA (earnings before depreciation, interest, taxation and amortisation—effectively the ongoing cash flow from the business before capital expenditure)', 'once annual revenues', 'net assets plus 50 % of annual revenues', or some similar measure.

In an industry that is consolidating into a smaller number of participants during a growth phase, a large number of businesses are frequently found which are backed by formal or informal venture capital investors, and which have ongoing requirements for cash as they develop. Most industries at an early stage go through varying waves of investor sentiment. The first consolidation usually occurs when sentiment to the industry has changed from one of unbridled optimism to one of harsh reality, and investor interest in committing new funds to the sector can dry up very quickly. In this situation, there may be large numbers of companies available for the well financed consolidator to acquire. There can be a very fine line between a business being worth nothing, as it teeters while running short of cash, and it being worth a significant amount of money if it is seen as having attractive characteristics for an acquirer.

An industry can sometimes see the emergence of category killers, such as seen in retailing. The most recent example was probably the video rental industry in the early 1990s where Blockbuster Entertainment and a number of other large well funded businesses were able to redefine the criteria for being a respected operator in this area. The appropriate store size was immediately redefined, and a number of marketing innovations were deployed which made it very difficult for individual operators to compete. These category killers can make smaller businesses, less well funded and having local rather than national scale, vulnerable to being seriously devalued in the minds of potential buyers.

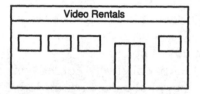

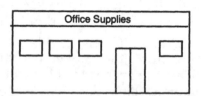

Superstores of typically 600 to 1000 m²
Competing mainly on:
● Ability to afford prime locations
● Greater scale of store allowing longer opening hours and broader range
● Ability to purchase as 'chain'
● Ability to afford press and TV advertising, better systems etc.

Suburban locations of up to 2500 m²
Competing mainly on:
● Appealing to new class of small-business buyer
● Greater scale of store allowing longer opening hours and broader range
● Ability to purchase as 'chain'
● Broad range of services, including software sales, copying etc.

Figure 8.1   Industries where category killers have emerged

8.3 Issues in the acquisition

Buying a business is always interesting. The homework, or due diligence, is done on the business and as much as possible learnt. On having made the acquisition, the perspective immediately changes. Up until now the focus on the part of people in the acquiring firm was on justifying the acquisition; now it is a question of making it work. A change in perception can occur; glasses that were previously seen as half full can now be seen as half empty, or possibly the reverse.

If an acquisition has been undertaken for financial reasons, without any expectations for synergy, and the existing management is remaining in place, then the business after acquisition may well continue on in a similar fashion. If management is changing—for example if the business is being bought from a retiring owner—then matters are different. And if the purchaser is seeking to change the culture of an organisation, then problems can arise. Frequently, for example, businesses are bought from the founding entrepreneurs. Usually it is very desirable that the founding entrepreneur stays, with the expectation that continuity can be maintained and that the drive and enthusiasm that was in place can be retained for the benefit of the new owners.

Securing continuity of service from a selling entrepreneur can be difficult. Usually the seller has become quite rich as a result of the process, which confers a measure of independence. The freedom enjoyed as an entrepreneur is not often willingly exchanged for a position, however senior, within the corporate structure of the acquirer. The most publicised example here is probably the General Motors acquisition of EDS, founded by Ross Perot, and the subsequent falling out between Ross Perot and General Motors. The transition may be eased by an earn-out arrangement, where the selling entrepreneur might be committed to the business for a period of typically three years after completion of the deal, with a material part of the consideration being deferred and based on performance during this period. This seems very logical, and often is. The disadvantages are

(i)   To achieve a fair measure of performance during the period, the selling entrepreneur is left largely in control of the business, as otherwise measurement of the earn-out performance becomes difficult.
(ii) Any measurement during a period may not be representative of the longer term health of the business.

The seller may be tempted to undertake actions which maximise reported profits or other measures during the period of the earn-out, which is fine, but may be reluctant to undertake the investment in marketing, technology and plant which is necessary for longer term development of the business, and which may subtract from reported earnings in the period in question.

In many cases, a corporate acquirer prefers to buy from another corporation than from a selling individual. If buying from another corporation the acquirer will know that the often difficult transition from entrepreneurial management or owner-management to professional management will have been undertaken, and may expect a corporate culture to be in place that is roughly similar to its own. On the other hand, if acquisition of a fast moving business is sought, the interesting targets are likely to be owner-managed businesses with little choice available.

The issue of corporate culture is important. Corporate culture subsumes many of the values and practices which exist within an organisation, and which make the behaviour of the corporation predictable to those working in it or dealing with it. This predictability is usually desirable for employees, even if too much predictability is not an advantage in the marketplace and can stifle innovation.

Key issues underlying corporate culture in these situations include

(i)   Employment policies—does the business aim to be a long-term employer or does it have a hire and fire reputation?

(ii)   Remuneration policies—are employees in key areas remunerated largely by salary or largely by commission or other short term incentive arrangements?

(iii)   Technology issues—some businesses pride themselves on the extent of the in house technical capabilities, while others have little respect for technology and see themselves as marketing businesses.

(iv)   Ethics issues—some businesses will see themselves as operating to very high ethical principles, while others may be more lax in such matters.

(v)   Some businesses will go to great lengths to inform employees about company performance, while others will seek to suppress this information.

(vi)   New style versus old style management. The new age management will probably have embraced concepts such as world class manufacturing, with its attendant emphasis on employee empowerment and on spreading broad responsibility concepts through the organisation. Old style management may be very much in the traditional authoritarian mode.

(vii)   Corporate culture issues dependent on differing customer focus. Most successful businesses are customer led, and will take on many of the characteristics of the customer industry which they service. In the electronics industry, for example, there has been limited interaction between firms using the same technology for different customer groups. The military market requires large overheads to be in place associated with formal quality checking for a relatively small number of units, built against long timescales and with a large emphasis on reliability and historically a lesser emphasis on price. Putting together a defence contractor and a consumer electronics maker, where the emphasis is on large quantity, low cost production at high quality but to a lower design specification, can give significant culture clash.

(viii) A business can also take on aspects of the culture of the geographical region in which it is located. Employees in California may be more relaxed on some matters than employees in the East Coast US. In southern Germany, beer may be readily sold within the workplace, although other countries have strong anti-alcohol codes. Acquisitions having a transnational character, or between companies from different regions within the same country, are prone to culture clash on these and similar grounds. Some issues in this context are treated in 'Executives are oceans apart on values' (*Wall Street Journal Europe*, 19 February, 1996).

Integrating management is often the desire. Management is usually retained, particularly if the business being acquired has been successful against criteria which matter to the acquirer. However, there can be situations where the acquirer wishes to replace management completely using its own nominees. This can occur if the acquirer has little confidence in the business being acquired, and is really seeking to acquire assets. It can also be the case where the company being acquired is at a transition. A company may be built to be acquired, for example. A successful formula used by some entrepreneurs is to develop the business to the stage where good technology is in place, a good production facility has been completed and a strong position is held in the local market. The entrepreneur and his team then sell and withdraw from the business, moving on to repeat the area at which they are skilled, namely developing businesses to a certain stage. The business can then be sold to a large multinational which will want to deploy its own management team.

Styles of acquirers will differ greatly. Some acquirers are quite happy to retain a federation of near independent businesses, (such as Dover Corporation), while others seek to integrate the businesses very completely, deriving synergies and seeking to reduce costs by having a common approach to accounting and other issues in order to minimise overheads.

Integrating the management team is not easy. Some companies value long service, so it is difficult for the incoming manager who sees his future blocked by many having longer service than he has. On the other hand, some of the longer servers can be threatened by the new arrival. Putting together management teams from different national backgrounds can be fertile ground for belief of discrimination. Sometimes the most difficult transition is when the company makes its first acquisition. Previously headquarters and the main plant were synonymous. Now the company has to realise that it is a multi-location company, and headquarters must be as equally accessible to the remote location as to the home location. This may mean that headquarters should separate itself out from the main plant, if only to allow the broader perspective appropriate to an incipient multi-location or multinational business!

On some issues the acquirer will want to impose uniformity, and in others will probably want to keep as much local autonomy as possible. Uniformity is needed on ethics issues, quality issues, financial reporting and in matters of logo, branding and other issues relevant to the corporate image. Beyond these, whatever approaches work best can be used. The acquired company that has a reputation for flair in new product design must retain this creativity and not be stifled more than is necessary by corporate controls.

Integrating information systems is usually one of the bigger issues associated with acquisitions. Given the plethora of software systems available for tasks such as accounting and materials requirements planning, an acquisition usually sees the need for two or more incompatible systems to start talking to one another. This is often the biggest barrier to success. The people will talk to one another, but getting the systems to talk is more difficult. In fast moving industries it is not realistic to stop and redesign from scratch. Illogical though it may seem, the best option is usually extending an old system, spending money on something that should be scrapped and replaced as soon as possible. These systems are essential to getting product out the door, itself a prerequisite for getting cash into the business and we cannot afford to be quite as decisive as we might wish!

Acquisitions of course do not always work—perhaps 50 % are judged as unsuccessful. The definition of work can vary, of course, very significantly depending on the viewpoint adopted. The acquisition may be successful in giving some of the non-financial returns, while being a failure in strict financial terms. Difficulties can be attributed to corporate culture differences, information system incompatibilities, different reward structures, lack of understanding of the key issues in the acquired business, etc.

8.3.1 Conflicts in an acquisition

In making an acquisition a company sends out signals about its business objectives. If it has acquired key distribution outlets, it may be competing with immediate customers. If it makes an acquisition of a supplier, it may be seen as competing with its current supply base. These conflicts can sometimes be managed or addressed in various ways, but other times they have to be lived with.

8.4 Designing the business to be acquired

An entrepreneur's skills may lie in building the business through its growth phase. When growth begins to taper off and business becomes more mundane, the entrepreneur may wish to sell out and repeat the process or retire. Also, the nature of the industry being addressed may be such that it is very difficult for a small firm to progress further without either acquiring (which may be beyond its capabilities) or being acquired. The business might have been designed *ab initio* to be a suitable acquisition candidate, recognising these industry characteristics. In this case the design objective must be getting the best value for the business on sale. Value may be defined as the entrepreneur wishes. It will usually have a strong financial component, but there may be social or prestige objectives in terms of having created employment in a region or being seen as a successful businessperson.

Designing a business to be acquired has risks. The first is of course that there may be a relatively small number of acquirers for the business in question. The period when maximum value is likely to be ascribed to businesses by acquirers trying to get their market share can be mistimed, or the wrong technology could be backed. The weaving pattern of alliances and emerging standards as an industry develops is not easy to foresee and a business may be left behind having been allied with the wrong camp or otherwise having chosen an unsuitable technological path.

In designing a business to be acquired, the entrepreneur must have regard to the management approaches of the acquirer. If the acquirer is just acquiring technology, then having developed a full management team may not be a major selling point. In other cases, the acquirer may be desirous of using the acquired vehicle as the means for entry into a new business area, and in this case retaining the management team can be crucial. In this case, the entrepreneur can add immense value by already having developed the management team.

A business unduly dependent on the entrepreneur has a reduced value in the eyes of the corporate acquirer. As noted, integrating an entrepreneur into a corporate structure is often difficult, and putting in professional management to replace the entrepreneur is a difficult task for the acquirer. The entrepreneur can maximise the position by leaving operating management of the business in the hands of a professional manager who is likely to adapt well to the post-acquisition environment.

8.5 Designing the business to be an acquirer

Changing circumstances, brought about by factors such as legislation or demographic change, can frequently force the pace of change in an industry, promoting the consolidation of numerous small businesses into a small number of large companies. The US waste industry went through a period of consolidation in the 1980s and the healthcare industry in that country is going through similar change in the mid-1990s. The growth of companies such as Novacare, consolidating rehabilitation service provision in nursing homes and outpatient service centres, or the growth of Apogee in the mental health area, indicate the trend here.

To be a consolidator requires a significant amount of funding, needed to buy businesses. The acquired businesses often have modest net assets, so much of what is being bought is goodwill. Borrowing against goodwill does not usually make sense, even if banks were to return to the aggressive lending practices of the late 1980s, so equity capital is required. The stock markets tend to be prepared to give a significant amount of cash at reasonable valuations to those companies which look as if they are about to consolidate an industry. The proving phase prior to the initial public offering involves demonstrating that a successful operation can be run in a small region, showing good operating profits at the local level, and that this concept can then be cloned nationwide or internationally.

8.6 Alternatives to acquisition

Acquisitions can be expensive when seen from the acquirer perspective. If the desired objective is access to resources, or achieving geographical coverage, how can this be achieved in some less costly fashion? Acquiring a business is often the no compromise answer, providing control of the business and its assets. These assets can be deployed as the acquirer wishes, and competitors can be denied access.

In the real world, the no compromise answer may not be affordable. A company may have to settle for other, less expensive, strategies to get resources, such as franchising and licensing, and alliances.

A firm may have a good local concept for a restaurant. Going for a public offering to expand the concept nationwide by acquiring or building restaurants locally to its formula may not be attractive either to investors or to the company. The solution may be to franchise the concept, collecting a royalty fee and benefitting from the hard work put in by franchisees. This does not bring the full control and value that would have been associated with full ownership, but dilution of interest associated with having raised money is not suffered.

Likewise, instead of buying out the technology for the emerging biotechnology company, it might be possible to get a licence from them for its use. Generally, the closer the licence is to being fully exclusive, which can be tantamount to buying the company, the more that must be paid. However, a limited exclusivity window could be accepted, recognising that getting the headstart over competition is often what is important. A good deal for this might be obtained from the biotechnology company, as they recognise that it does not threaten their long term independence.

If geographic coverage cannot be achieved as a service provider to the computer industry, partnerships might be developed internationally. The important aspect here is keeping these partnerships real and giving a seamless interface to the customers, similar to what they would see if dealing with a single multi-continent business.

Alliances might be formed, underpinned by equity investment. Strategic alliances of this type have been in vogue in banking and in the airline industry, and the telecommunications industry is also seeing their emergence.

It can be questionable what a minority stake really does in these situations. In the minds of all participants, it is a declaration of mutual commitment and it makes it difficult for the company selling the minority stake to choose another partner subsequently.

8.7 Paying for the acquisition

In funding an acquisition a wide range of financial instruments can be used. The simplest approach may involve a cash payment to the vendors. A public company may make a paper offer, issuing its own shares to the vendors. If the business is to be paid for directly, this cash must be raised. There may be a rights issue to current shareholders or debt facilities may be employed. If the business being acquired has significant tangible net assets, then these could provide additional borrowing capability. On the other hand, if a business with minimal tangible net assets is being acquired, with a high figure for goodwill or other intangibles, then the overall borrowing capacity is reduced. Non-recourse funding would possibly be available for the assets of the business, where the acquisition funding can effectively be ring-fenced from the funding of the overall group. These techniques have been used by property companies and by mining/resource companies, where the bank has a clear charge on the assets in a development and no or limited recourse against the parent company.

Section V
Management buy-ins and buyouts

Chapter 9
Management buy-ins and buyouts

A very large part of the available pool of venture capital funding in recent years has gone towards purchase of businesses by their managers (management buyouts), or to purchase of businesses by managers with an established track record (management buy-ins). There have also been various hybrid arrangements where existing management has been augmented by incoming management with specific experience complementing the existing team. The term BIMBO (buy-in management buyout) has become common in the industry to describe the full range of transactions of these types.

9.1 MBO considerations

Why are so many existing owners of businesses keen to sell to management teams, backed by money committed by venture capital sources? Motivations are usually as follows

- the vendor may well get best pricing from a well backed management team. This might seem counter intuitive, in that a trade acquirer could take on the business and achieve efficiencies that would justify their being able to pay a higher price than the backers of the management team could justify, and they will gain no synergies from the deal. Financial buyers, backing management teams, may, however, be able to undertake some financial engineering to allow them to pay a price competitive with trade buyers
- the sale can be to a non competitor, with the opportunity in some cases for the vendor to retain a small continuing interest in the business

- a developing preference for focusing on core competences may motivate sale of internal service businesses, perhaps with a guarantee of some continuity of business. Sale to management may be the easy option in these cases
- the sale can be communicated in a very positive light to community interests in the area in which plants are located, and it can be communicated as socially acceptable
- the transaction can also be sold very well to customers, who otherwise might worry about being abandoned by the larger group
- particularly if the vendor is an individual, there may well be an aspect of seeking to have the business remain in good hands. An individual may well be allowed to have a greater sense of community identification than in the case of a company with shareholders, and may well place a high value on standing in the local community. Selling the business to management may enhance such standing. The retiring founder can also arrange for a gradual disposal over time to the management team.

Particular attractions for the investor are as follows

- there is the opportunity of investing in relatively large businesses with established market position and reputation
- businesses usually have in place all the controls and financial information appropriate to a large group
- managers in large groups are sometimes perceived as not having freedom to manage their particular operation to best effect, or their operations may be starved of investment. The belief may exist that management could do far better given focus, freedom and extra motivation
- buying established businesses with (hopefully) predictable earnings and cash flow characteristics, and with some asset backing, may allow the use of significant amounts of debt, which means that the amount of own funds committed by the investor can be relatively modest
- there is usually an opportunity to achieve a relatively rapid and significant gain.

The types of businesses usually preferred by investors are thus

Businesses with a clear exit mechanism

The investor in such transactions is looking to buy the business for today's price, and be able to sell it on at a significantly higher price within perhaps three to five years. Businesses that have the scale to go public on the stock market, or which might be attractive candidates for selling on to a trade buyer after a period of independence, are usually the types of businesses favoured. Or, there can be a situation where periodically the businesses are refinanced again, with new investors taking the place of those who may need to exit. Several companies have gone through successive management buyout transactions.

Independent businesses

This ties in with the exit mechanism issue. It is difficult to undertake a buyout of a manufacturing plant or a service centre, without clear visibility as to how it would obtain new business in the near term in order to become largely independent of its current parent (perhaps with the objective of having sales to current parent less than 25 % of total sales after three years).

Businesses with stability and predictability of profits and cash flow

Some modest levels of growth may be desirable, but the main focus is on ensuring that the business will perform with a high degree of certainty at least at current levels of profitability and cash generation. This allows significant levels of debt funding to be raised, which can increase significantly the potential upside available to the funder.

Businesses that can tolerate a capital expenditure holiday

The objective of the MBO exercise is to achieve maximum increase in value for the investors, with minimal commitment of capital. Ongoing maintenance capital investment will be needed, but it is clearly a bonus if major plant investment has been undertaken by the parent company just before the buyout proposal is developed!

Businesses with opportunities for increase in profitability in the near term

An independent and motivated management team can usually reduce costs in a business very significantly, being more open to new ideas for re-engineering the business. The aim, however, is to sell on the business for significantly more than was paid for it, so cutting brand promotion and product development activities may well be counterproductive.

Businesses which have a broad strength in the management team

In a management buyout, the focus is on avoidance of negative surprises, and with the expectation of some upside within a three year period. All the key bases need to be covered by the management team, or new management brought in to fill in areas where there may be gaps. The managers sought are those who are good at operating issues and at getting results in a relatively short timeframe. Having strength in depth in the team also means that the business is less dependent on one individual, which can be a particular point of risk.

Further discussion of management buyouts, including many interesting anecdotes from those who have been through the process, is presented in NASH, 1995.

9.2 Financing aspects associated with management buyout transactions

The MBO or BIMBO business has become quite developed in recent years in most developed countries, and thus has its established transaction structures and procedures.

The issues involved are best understood by working through an example, which is broadly representative of current practice but which does not bear any intentional resemblance to any individual transaction.

9.2.1 The genesis of the MBO

The directors of General Consolidated (GC) were considering the future of their cableform manufacturing operation (CMO). This business had been acquired when diversification was more popular than currently, and it had no links with the rest of the GC businesses. CMO did a lot of work for customers in the personal computer industry, and GC directors were uncomfortable with the long term opportunities for making money from this sector.

Key financial figures for CMO were as follows, for the year just ended and for the four previous years

Year	−4	−3	−2	−1	just ended
Sales (£M)	125.0	130.0	135.0	132.0	130.0
Operating profit (£M)	3.5	3.5	4.0	3.0	2.0
Net fixed assets (£M)	20.0	17.0	14.0	11.0	9.0
Working capital (£M)	12.0	12.0	12.0	13.5	15.0

Viewed from the GC perspective, these figures showed

- poor growth in sales over the five year period, with slippage in recent years
- profits erratic, with most recent figures half of the peak performance two years earlier
- working capital in the business was also tending to increase. Combined with declining profitability, this was not encouraging. Making £2 million operating profit on assets (net fixed assets plus net working capital) of £24 million was not a good return by any criteria.

GC had been coming under pressure to focus more clearly on areas perceived as being closer to its core business. Given this pressure and the CMO results, the decision was quite easy. 'Divest of CMO' was the clear objective as expressed in the minutes of the relevant meeting.

Joe Smith was charged with selling CMO. Looking at the numbers, he knew the case was quite clear. He had also, however, built up some rapport over time with Mike Davy, the manager of CMO.

Joe Smith could accomplish the sale process in a number of ways. He could put the business up for sale very openly, he could doorknock competitors for CMO, or he could offer it to management. Mindful of his fiduciary obligations to the GC shareholders, he decided that any process had to be transparent. He went along to see Mike Davy at CMO to give him the news.

The understanding developed with Mike Davy was that there would be something similar to an auction, but that management would be given the opportunity to match the leading external bid for the business. This seemed to satisfy all objectives. It also removed Mike Davy from some of the ethical issues involved here. He, and a small number of trusted colleagues, agreed with Joe Smith that they would be quite open with prospective purchasers, but would have this chance of matching the winning external bid. Under conditions of strict confidence, the management team was also allowed to make contact with two MBO funders.

In an MBO situation, management is always in a position of divided loyalty. It has a responsibility to the selling company to get the highest price, and likewise it wants to be able to buy the company at the lowest price. Joe Smith reckoned that the approach adopted, while not eliminating some of the inherent conflict of interest, would at least seem fair and not inconsistent with getting the best result for GC.

Joe Smith thus produced the information memorandum on CMO. In just over 200 pages, it described the business, outlined the performance in the last five years, gave details on management, plant locations, equipment etc. This was provided to parties who had expressed interest and who had signed a confidentiality agreement, following publicising of the opportunity on a no names basis through various investment banks and business brokers. Joe also advised key customers of CMO, on a courtesy basis, not least because some of the CMO customers were important to GC in other areas.

These things take time, and eventually four offers came into place for CMO. One was from a rival located adjacent to the main CMO facility. The acquisition made very good synergistic sense for the acquirer, but the price did not really reflect this. Joe was also a little apprehensive of the reaction of some of the customers to this news, given some problems earlier at another division of this suitor. Another offer was for a higher sum, but with the stipulation that GC take responsibility for taking over all the management team. Joe could not see a way in which the team could easily be integrated within the rest of GC, and thus there was a high implicit cost to be set off against this offer. Another offer was for just part of the business, but Joe reckoned that the remaining part would not be viable and closure costs would be high. The top offer, in Joe's mind, was one from a US company in the same business as CMO. It fitted well in terms of customer continuity and, given that the proposer did not have any operations in Europe, it looked as if it could be very acceptable to the members of the management team, who would be likely to receive higher salaries and status, along with having a more supportive parent.

The proposed deal valued the business at £25 million, payable in cash. Joe thought this deal was very good. Applying a notional tax charge of, say, 33 % to the operating profit of £2 million, this was a P/E of 19, and there was a very modest gain, but a gain, in book value terms. He felt that he could justify this well within GC. All sounded fine.

Mike Davy was thus presented with a more severe hurdle than he had earlier envisaged. If he wanted to buy the business, he and his team had to raise £25 million. And they had to get it within a month!

Thankfully, he had been permitted to have some initial discussions with backers.

The story put over by Mike Davy to his potential backers was essentially as follows

(i)   GC did not understand CMO. This was evident in under-investment in the business in recent years, particularly as indicated by the falling figures for net fixed assets. The CMO plant was literally starved of investment, and was thus much less efficient than it might be.

(ii)   CMO's poor working capital position was partly as a result of GC's policies. These led CMO to over order on stock, in spite of which production was often delayed due to material shortages. GC followed a policy of stalling payment to vendors, which was having a counter productive effect at CMO.

(iii)   The key point was that CMO had in recent times moved a lot of its business away from mass market automotive and PC businesses into specialised vehicle businesses and particularly telecommunications. Mentioning telecommunications to the investors caused much more interest. This could now be positioned well for exit as a supplier to a growth business. Margins were also improving, and Mike Davy was able to put forward soundly argued projections based on modest capital investment and on a gradual move into the higher growth and higher margin areas. These showed operating profits growing to £7 million, on roughly similar turnover, in the next three years.

Mike Davy actually spent a lot of time with his two potential funders. In the end choice of one funder was very difficult. Given the size of transaction, both were reluctant to be involved even in a two investor syndicate, but in the end they agreed. If after tax profits could be about £5 million after three years, and the company positioned in growth sectors, then a valuation of £100 million would be quite realistic. That sort of upside was suitably tempting! The investors were also aware of a potential candidate for the role of marketing director of CMO. This individual was invited by Mike Davy to join the team, thus giving an aspect of management buy-in as well as management buyout to the operation.

When the news broke to the investors that £25 million was the figure to beat, they were not too surprised. There was always the hope of getting a cheap deal, but that is a rarity nowadays. However, a huge amount needed to be done in the 30 days outstanding.

The first task was to work out the components of the financing. The business here was being bought at close to net asset value, so a lot of bank debt could probably be used. On the other hand, a further investment program was required. Some of this might be funded from savings in working capital, but further dedicated funds might be required for this purpose.

A first cut analysis might suggest that the total business funding required would be £30 million. This would essentially be the £25 million purchase price, £1 million in expenses and £4 million in additional funding, to cater for capital expenditure plans and additional working capital requirements. The bank might be prepared to provide £16 million at, perhaps, 8.5 % interest, and could give more, albeit at higher rates. Management might show commitment with investment of £300 000, and this would leave a requirement for funding by the investors of about £14.7 million jointly. This might be about £2.7 million in pure equity, and the balance in subordinated loans at perhaps an interest rate of six per cent.

The funding structure outlined had then to be reconciled with the reality of cash flows from the business.

Projections prepared by Mike Davy for his backers showed operating performance of the business as follows

Year	current	+1	+2
Sales £M	130	130	130
Operating profit £M	2.5	4.5	7.0
Cash flow (before capital expenditure)	4.0	8.0	11.0
Capital expenditure	3.0	4.0	4.0
Cash flow (pretax) to lenders/investors	1.0	4.0	7.0

These figures showed that in the first year there was in fact almost no net cash flow, but this had been allowed for in any event with extra funds being supplied for the purpose of capital investment. Allowing for this, £4 million would be available to meet requirements of lenders, and the figures in subsequent years are again £4 million and then £7 million. Eight and a half per cent interest on £16 million works out at £1.36 million, and the subordinated debt from the investors comes out at £720,000, involving a total annual commitment of £2.08 million, which can be readily accommodated within reason. It is assumed that repayment of the principal is held off until at least into the third year. It might be possible to employ more aggressive gearing approaches, but this structure would appear sensible.

The key role in the transaction was of course being undertaken by Mike Davy. He was investing, together with his team, a total of £300 000. This was not a large sum in the context of the overall funding, but obviously quite material for the team involved. The structure as worked out initially by the investors would see this amount buy ten per cent of the new company, which would take over the business and operations of CMO. Negotiations with the investors finally meant that he also secured options for the management team to acquire shares at a similar price to their buy-in price, which would give them an effective 20 % interest in the business.

In this situation, the favourable scenario is where profitability and cash-flow forecasts are met, and then after three years the business is ready to be floated. At this stage an operating profit of £7 million has been achieved. The business is still in debt, but good cash flow has allowed bank debt to be reduced to £10 million, and with this sort of operating profit the business is valued at £90 million. Raising some new money in the initial offering to pay down debt (bank debt and subordinated debt from funders) means that the value of the equity is now £65 million, a good return on the figure of £3 million committed earlier. The management stake is now worth approximately £12.7 million(entitlement to 20 % of the business less the cost of exercising options), and the funders have got a return of £52 million on the £2.7 million that they subscribed. Their return is naturally to be computed based on the overall funding they provided, including the subordinated debt, and in this case an investment of £14.7 million has become worth £64 million, plus the interest on the subordinated instruments. Assuming a total transaction time of three and a half years, then the internal rate of return works out at approximately 52 %, generally considered to be very acceptable!

9.2.2 Issues involved and alternative outcomes

Alternative company approaches to MBOs

The first point is that in this situation GC really adopted quite an enlightened approach. In other cases it might have decided to put the unit up for sale, excluding possible management involvement. This might be rationalised based on a fear of management being uncooperative and complicating the sale process, if they were advised at an early stage. As noted, there is usually a fine line between management loyalties and ethics where duty to an employer has to be balanced with being able to take on the opportunity of running a business.

Backers might also not have been convinced so easily about the nature of the opportunity, and the ease with which Mike Davy would be able to improve performance. To the casual observer, the history of the business expressed in financial terms was anything but attractive, and GC appeared to have a good deal lined up.

Companies will naturally vary in their policies on MBO transactions. If a large company wants to maintain uniformity across a range of businesses, then encouraging MBO transactions might well open up a Pandora's box of difficulties. As with other areas of corporate endeavour, there are also fears of less than ethical conduct in the case of management buyout transactions, with management having the opportunity to depress the performance of a business in order to motivate corporate management to sell it at a cheaper price than otherwise might be the case.

Bank funding issues

In the example given above, the proposition is probably more bankable in the conventional sense than many management buyout transactions. In many cases there might be much less asset cover, but the cash flow from the business might be excellent. One can often have in a management buyout transaction the situation where the acquiring company has negative net worth. This means the balance sheet might appear as follows

	£ million
Net fixed assets	9
Net working capital	_3
Total	12
Funded by	
Debt	15
Shareholders' funds	(3)
Total	12

Such a situation usually arises where a significant sum is paid for a business with a very small tangible asset base. The resulting surplus paid over the tangible base—the goodwill—is written off, thus giving a deficit in shareholders' funds. A banker lending into this situation naturally cannot take comfort from having a large amount of assets on which the loan can be secured, but on the other hand the ongoing cash flow from the business may be more than adequate to pay interest and to convince him that the loan is safe.

9.2.3 A successful MBO experience

One successful MBO in the electronics industry was the transaction concerning Vero Group plc (Vero). Vero was founded in 1961, with the initial product being the Veroboard prototyping system. The business then developed into other areas of circuit broads and into power supplies and enclosures, before being acquired by BICC plc in 1979. BICC continued to develop the business, including undertaking the 1986 acquisition of Imhof-Bedco, an established manufacturer of racking systems for telecommunications applications.

The Vero example is relevant in showing the gains that can be made by managers, employees and investors in a buyout transaction, and in illustrating some issues associated with the flotation of such a business.

In the early 1990s, BICC sought to concentrate on its core cables related businesses, and the Vero operations, being considered non core, were made available for sale. The management buyout team, led by Brian Gay, chief executive, acquired the business with backing from buyout funds Candover and Mercury Development Capital. The business was bought in April 1994 for a total consideration of £20.9 million, with £1.8 million of this being costs associated with the transaction. Taking in assumed debt, the consideration was £35 million.

Vero, operating as a division within BICC, had no statutory accounts, but management accounts proved sufficient for deriving appropriate figures for the business elements that were bought by the new company, Vero Electronics Group Limited, later with the name changed to Vero Group plc (Vero). The performance of Vero for periods before and after the buyout, as stated in the prospectus in October 1995, was as follows, with a profit forecast being given for the full year to 31 December 1995

	Year ended 31 December			Six months ended 30 June	
Year	1992	1993	1994	1995	1995
Turnover (£ million)	61.2	65.8	78.7	47.6	96.0
Operating profit (£ million)	2.2	3.3	8.3	7.0	12.6*

*The full year operating profit was computed before an exception item relating to an employee bonus payment payable on flotation.

Particularly evident from these numbers was the success of management in driving forward growth in sales, from £65.8 million in 1993 to £96.0 million in 1995 (46 % increase). This growth was led by strong demand from telecommunications customers in particular, with cellular base stations, requiring considerable amounts of racking, cabling and power supplies, showing particularly strong growth in this period. Even more impressive was operating profit growth, from £3.3 million to £12.6 million, a growth of 280 %, reflecting operating margins going from five per cent in 1993 to 13.2 % in 1995. These figures presumably reflect the results of increased investment, and reflect the use of modern manufacturing techniques, including manufacturing cells, kanban stock control systems, just in time methodology and CAD/CAM (computer aided design and manufacturing) processes.

The assets of the business being acquired on 14 April 1994, on a fair value basis, were made up of

£ million		
Tangible fixed assets		8.6
Stocks	10.4	
Debtors	14.9	
Cash	6.3	
Current assets		31.6
Total assets		40.2
Creditors due within one year		13.0
Creditors due after more than one year		16.9
		29.9
Net assets		10.3

The price paid to BICC was £19.1 million in cash, with costs bringing the total figure to the £20.9 million. The excess of this figure over the net assets of £10.3 million is really a goodwill figure, in this case of £10.6 million. Also, debt of the acquired companies of £14.2 million was assumed, resulting in a gross consideration of £35.1 million.

The amount of £20.9 million was funded by

£ million	
Unsecured subordinated loan notes	14.5
Other loans (vendor)	2.0
Share capital and share premium	2.0
Bank loans	2.4
	20.9

Total funding subscribed for A shares, by members of the management team was £369 000 of a total of £1 969 000, for an interest of 18.7 %.

Vero was floated on the stock exchange in November 1995 at a price of 210p per share, with £19.4 million being raised for the company, net of expenses. The original equity stake acquired by management was now worth approximately £16 million.

9.3 Management buy-in/buyout—the future

Buyout and buy-in opportunities are likely to be with us for some time, as firms continue to shed non core activities, and seek to outsource operations to former captive units. The large number of family businesses in Europe, founded in the immediate post war years, has also been held as presenting an interesting opportunity for management buyout and management buy-in transactions.

Management buy-in and buyout transactions are probably also very good for industrial regeneration, in that they achieve a closer link between management and ownership of capital, a link which can become somewhat tenuous in the case of larger corporations. Buy-in transactions also present an opportunity for revitalising firms which may have gone stale under earlier management and ownership structures.

From the entrepreneur's viewpoint, it is obviously a lot easier to undertake a management buyout than to start up a new operation. An existing profitable business paying the salaries of all concerned can also present a good jumping off point for new opportunities which may arise. From the funder's viewpoint, there is the opportunity to put a significant amount of money into an established business, with a good outcome very likely. Given the existence of competitive venture capital markets, the attractiveness of the MBO transaction has led to competition between funders, in many cases based on price. A natural consequence of paying highly to buy into the business is reduced return to the funders, which may result in their trying increasingly leveraged structures with higher risk, or possible whittling away of management incentives.

The management buy-in is less straightforward from the viewpoint both of the venture funder and the manager. The manager buying-in does not know the business in detail, and may or may not fit with the existing team. There is thus a greater element of risk for the funder and for the manager, but on the other hand MBI transactions may be smaller and better priced to reflect this situation.

Section VI
Conclusions

Chapter 10
Commercial judgment, leadership and other people matters

10.1 Knowing the game

In the commercial world, there is a constantly competitive environment. The game is ongoing, without the predictability expected from physical systems. Being able to play the game effectively requires very good judgment, under conditions of limited knowledge.

Every business has its own rules, sometimes to be observed and sometimes to be consciously broken, but always to be known. It may be convenient to go with the conventional wisdom in an industry concerning margin structures and commissions, or elect to come in as the innovator in this area and upset long standing arrangements, but the consequences of actions must be known. Every business area has its stars, those who continue to do better than other companies, with the performance usually being accounted for by numerous areas of attention to detail. Hence, when entering an established industry, investors will usually place a premium on the detailed commercial industry knowledge that it likely to be possessed by the executive who has had successful profit and loss responsibility for a division or a business in the area. When entering a new business or seeking to upset the applecart in an established business, then of course the skill set can be quite different. The office superstore industry in the United States was, for example, started by executives from the supermarket industry, using these skills to revolutionise an established industry with new business models.

Judging how to price the product or service is important. By understanding buyer behaviour well, proposals can be designed so that the buyer will be prepared to allow a higher price and still give the business. Developing the discipline of turning away business because it is not correctly priced can avoid development of profitless prosperity. In some cases differential pricing structures can be developed, such as the airlines employ to fill seats, burdening cheaper tickets with ever more onerous restrictions in order to segment the market and get maximum revenue from each type of traveller. In tender or bid situations where price is the key determinant of selection, the objective is naturally for the seller to come in £1 below the next lowest offer, and be confident that satisfactory profit can be made on the transaction at this price! Judging pricing takes, once again, real knowledge of the buyer's mindset, and also good judgment.

Whatever the game, it is important also to establish a reputation for honesty and for the ability to be trusted. It is important to keep to commitments, and if they cannot be met for some reason, to be honest as to the cause.

10.2 Judging people

Business is fundamentally about people. It is people who buy from and who work for a company, and most key business interactions are on a person to person basis. Understanding people, being able to recruit them, lead them and negotiate with them represent key business skills. There is an insight that can come with experience which allows a person to determine on whom to rely and of whom to be wary.

Judging people who will make up the team is important—many venture capitalists will claim that it is their key skill. They have to back management teams who will make money for both them and their investors, and being able to pick out the competent from the mediocre, or the exceptional leader from the average is much of what the venture capitalist's workload is all about. In recruiting the team, the entrepreneur needs to use good judgment, or to supplement his own judgment with inputs from an experienced professional. Having a good sounding board such as a capable non executive director can be of particular value here.

In recruiting, it is important for the entrepreneur to resist the temptation both to hire someone similar or not to hire someone with better qualifications. Most managers feel comfortable with those with roughly similar backgrounds, and the manager can feel threatened if someone with more experience is recruited. These approaches lead to the team being imbalanced, usually being biased towards one functional area and deficient in others. If a world level company is to be created, the best people that can be afforded must be recruited. The ambitious entrepreneur is going into business expecting most of the rewards to come from owning a material stake in the business rather than from the salary paid for an executive role, and so is the main person being cheated by a reluctance to hire good people on grounds of possible rivalry.

10.3 Leadership

Leadership has a very important role in companies and can come in many forms. Either charismatic or in the form of a more aloof respect, leadership is all about getting people in the organisation to feel good about common goals, and to give of their best. The difference between organisations that are led and those that are merely managed is very striking. People like to work for an organisation which has a clear sense of purpose, and a clear mission, if they are to devote a large part of their working hours to the company in question, and this mission is often articulated and communicated particularly well by the charismatic leader. A good leader also sets an example, in terms of commitment to the business and in communicating the values and culture of the company.

10.4 Selling

The very act of trying to sell something can come as a culture shock to many who have had a more commercially sheltered existence in technical areas. People are often particularly reluctant to intrude on someone else, to make the cold call, lift up the telephone or send out the letter. Salespeople get rejected, and no one likes to be rejected, which can explain the reluctance which may be felt. A more positive view has to be taken of the situation!

The attitude has to be that the potential customer is receiving a favour by being introduced to a product or service. There has to be an enthusiasm about both the company and the offering that the customer can sense. The customer needs to be shown how money can be saved, quality improved, or other benefits gained as a result of involvement with the company; the good salesman is really a counsellor for facilitating the decision. To sell properly the customer should be known as well as possible, and the approach made to the key buyer or influencer. There are numerous works published on sales techniques, almost all of which have useful insights into this important area of human psychology!

In areas led by people who are technically qualified, there can sometimes be the perception that selling is a less than worthy activity, and that if a better mousetrap is devised, customers will flock to buy it. This is, of course, not the case. Selling is ultimately about communicating, and the advantages of the better product will not be apparent, at least not to the extent of paying more for it, unless the message is communicated well to the customer. In many businesses where there is very little differentiation between the offerings from competing companies, the relative excellence of the sales force can be the key determinant of success. Many businesses have a strong 'hustle factor', where the most important element for success is to have the products better and more energetically sold by a committed and incentive led salesforce.

Selling is of course an activity that has to be backed up by a clear understanding of the buyer behaviour issues and by a clear marketing orientation of the business. There is often the complaint that the salespeople always sells 'what we don't have', but the blame for this can rest either on the salesteam or on poor specification of the product in the first place!

10.5 Negotiation

Much of business also is about negotiation. Negotiation can sometimes be turned into a win-win result, for example when seeking to put together an alliance for mutual benefit. There can also be an honest exploration of the issues that are important to each party, and those significant to one party but trivial to the other can be traded.

In many cases negotiations come down to a straight zero sum game, at least in the short term, where what one company gains comes directly from the opponent. Knowing when to take a stand in order to win, in the process putting many aspects of the relationship at risk, and when to accept the inevitable graciously, are key issues which depend on the judgment and personality of the negotiator. Getting to know the opponent, and determining subtly how far they are really prepared to go—what the bottom line really is—is key to the artform of the successful negotiator.

The good negotiator will also use the element of time, seeking to win points when the opponent feels under time pressure with a deadline looming or with a flight to catch. The only response to this often is patience, or at least informing the adversary that the tactics being employed have been noted! Good negotiation also means not acknowledging the weakness of a position, and playing always as if from strength, whether this be the case or not. The poker face approach, ensuring that the game is not given away through body language, is a skill to be cultivated.

A hard but fair negotiator wins for the company, but also wins respect for himself. Both sides gain satisfaction from a hard-negotiated transaction, while the soft deal invites suspicion, one side thinking that they got too little and one suspicious that they could have got more from the transaction.

The venture capitalist will often negotiate the terms of the deal very carefully with the potential entrepreneur, the negotiations being a test of how the entrepreneur is likely to negotiate with customers and suppliers. Selling equity in a business is one of the more important sales tasks to be undertaken, and the investor will take little comfort from a soft negotiation performance from a prospective client.

10.6 Attitudes to risk

Business is all about taking measured risks under uncertainty. No businessperson gets all decisions right and the most experienced managers can stumble. The approach is to contain the risk, or if there is no choice and it is necessary to bet the company, at least the management should know what it is taking on and ensure that it gets full commitment of all involved. Betting the company intentionally with the full knowledge of all is one thing, but doing so accidentally is one of the most serious business errors!

An approach often suggested is to experiment, but to cut experiments very quickly if they appear not to be yielding results. All businesses have problems of some sort from time to time, and it can be appropriate to establish a separate workout team to address the problem issue without distracting the whole organisation from the development agenda to be followed.

10.7 Contracts

Every business will have occasion to enter into contracts of various types. These can include employment agreements, lending and other funding agreements, intellectual property agreements (e.g. related to patents and licencing issues), agreements with distributors, suppliers and customers, confidentiality agreements and premises related agreements. Such agreements are intended to bind the company for some period into the future, and none of these can be treated lightly. Agreements in the form of a simple letter can be as important in their consequences for the company as a long volume with formal legal documentation.

It is common sense to read all these documents carefully and understand the implications. Lease agreements on premises can mean that there is a liability for a number of years of rental, and the unscrupulous will often insert traps for the unwary in documents. In the western world, there is a somewhat litigious environment, but it makes sense in any event to get legal advice if there is anything that is not understand.

In preparing contracts, it is usual for the parties to prepare a heads of agreement or term sheet which sets out the basic points involved, and which is usually non binding except possibly for some points relating to confidentiality of information supplied, or to a period of exclusivity that may be available to one party in order to complete the transaction. This outline is usually then given to lawyers to put in more formal language which seeks to remove some of the ambiguities that can characterise a brief list of points. The important thing here is that negotiations remain on a principle to principle basis, without the lawyers becoming the scapegoats for disagreement on details.

It is also important to have procedures within any organisation so that the organisation cannot be bound by an onerous contract without due consideration. Most organisations will seek to have a requirement of board approval for any contract which could involve exposures above defined limits. Such arrangements are relatively easy to enforce in the case of capital expenditure and recruitment commitments, but are less easy to police with ongoing, routine trading. This can be a major risk area for many companies, where the company can be bound effectively by the actions, either inadvertent or fraudulent, of a low level employee.

A particular risk in the case of technology companies lies in taking on fixed price design work. In the case of software projects, such contacts can easily get out of control. Poor initial specification can result in a faulty design, and then as the deadline approaches, more people may be thrown at the problem, which if anything may slow down the process, all the time adding to cost. In other cases, a company may commit itself to specifications which are technically impossible or cannot be met due to patent issues, with the client taking action for losses incurred due to delivery failure and the design company suffering from having committed large resources to the project with nothing to show for the work. Careful attention to the provisions of contracts and their possible significance is important if a company is not to find itself inadvertently taking on risks of this type.

10.8 Interpersonal skills

The most important business skills are often in fact those interpersonal skills of negotiation, leadership and issues involving judgment, that are most difficult to communicate in written form! A good work on these topics is that by Mark McCormack, chief executive of IMG, best known for its management of sporting personalities. (MCCORMACK, 1984).

Chapter 11
In review

This, then, has been a tour of the business landscape, seeking to communicate the main issues associated with business to technical professionals, specifically engineers. The focus has naturally been on entrepreneurs and management teams, but much of the analysis and presentation is also relevant to those having to manage business units in larger corporations. In this case the issues are often the same, but the upside and downside associated with actual ownership of the business are not there, except in a contained form expressed by rewards in the form of bonuses and other career advancement issues.

The contention remains that engineers are well placed to understand business issues. Engineers in particular have to have an aptitude for design, and these ingrained design skills and capability can be put to good advantage in working with key commercial issues, as treated in the initial chapters of this work.

Finance is based on quite basic systems concepts, but naturally cloaked in a language associated with the professions that work with finance on an ongoing basis, primarily the accountancy profession. Understanding some of the language and concepts involved, and understanding the use of financial concepts in business, is central, and this has been discussed in Chapter 5.

Any consideration of financial issues indicates very quickly that a new or developing business is likely to have considerable appetite for investment, and the sources available have been outlined in some detail in Chapter 6.

Alliances have become particularly important to technology companies. Companies such as Microsoft have prospered particularly well from alliances, and alliances continue to be formed as companies need to enter new markets with good chances of success. The move towards outsourcing has led to supply arrangements that have the character of alliances. Alliances deserve to be taken seriously, as correctly done they can leverage the market access and valuation of the company very significantly. Done poorly, it is easy to 'give away the crown jewels' of the company, and the consequence can be sale from a weak position some years later.

Chapter 8 focuses on acquisitions. Some high technology companies will be acquirers, and many more will be acquired as industries consolidate. Understanding the acquisition process and key issues involved is essential for those on both sides of the transaction, in that the management team of the business being sold must understand the motives of the acquirer if it is to secure the best valuation, and the management team of the acquiring business must be aware of both alternatives to acquisition and the valuation issues involved.

Management buyout and buy-in transactions are popular, and are appropriate given the willingness of companies to sell off non core units and to move captive units providing services into independent operations providing the same service to a broader range of customers, and thus achieving greater economies and having greater encouragement to develop than if dedicated to a single customer. Such transactions are also appropriate given the pool of experienced, rounded managers that is developing, and it is in the interests of industrial regeneration that these managers be well placed to have an element of ownership of the businesses in which they are involved.

Finally, it must be noted that the necessary business skill set also contains a lot of items that are not communicated easily in written form. Judgment, negotiation and leadership are central to business success.

Bibliography

ANSLINGER, P. L. (1996): 'Growth through acquisitions: a fresh look', *Harvard Business Review*, January–February, pp. 126–135

ARUNDALE, K. (1995): 'A guide to venture capital' (British Venture Capital Association)

ARTHUR ANDERSEN (1993): 'Best practices' (Arthur Andersen)

ASIMOW, M. (1963): 'Introduction to design' (Prentice Hall)

BAKER, F. T. (1972): 'Chief programmer team management of production programming', *IBM Systems Journal*, (1), pp. 56–73

BELL, C. G. with MCNAMARA, J. E. (1991): 'High-tech ventures, the guide for entrepreneurial success' (Addison Wesley)

BHIDE, A. (1992): 'Bootstrap finance: the art of start-ups', *Harvard Business Review*, November–December, pp. 109–117

BHIDE, A. (1994): 'How entrepreneurs craft strategies that work', *Harvard Business Review*, March–April, pp. 150–161

BLEEKE, J. and ERNST, D. (1995): 'Is your strategic alliance really a sale?', *Harvard Business Review*, January–February, pp. 97–105

BROOKS, F. P. (1974): Datamation, reprinted *in* WASSERMAN, A. I. and FREEMAN, P. (1976): 'Tutorial on software design techniques' (IEEE Computer Society, Catalog No. 76CH1145–2 C, 1976)

BUZZELL, R. D. and GALE, B. T. (1987), 'The PIMS (profit impact of market strategy) principles—linking strategy to performance' (The Free Press)

BYRNE, J. A. (1993): 'The horizontal corporation', *Business Week*, December 20, pp. 44–49

COOPER, R. (1996): 'Control tomorrow's costs through today's designs', *Harvard Business Review*, January–February, pp. 88–97

DEAL, T. and KENNEDY, A. (1982): 'Corporate cultures—the rites and rituals of corporate life' (Addison Wesley)

DRI EUROPE (1994): 'Sectoral patterns in strategic alliances' *in* Panorama of EU industry—1994 (EU Publications Office, Luxembourg)

EARLY, L. (1994): 'The 'methods' of a quality master—an interview with Genichi Taguchi, father of quality engineering', *McKinsey Quarterly*, (4), pp. 3–17

'Evca yearbook 1995' (European Venture Capital Association, Zaventem, Belgium)

'Executives are oceans apart on values', *Wall Street Journal Europe*, 19 February 1996.

FOSTER, R. N. (1986): 'Innovation—the attacker's advantage' (Macmillan London)

FREEMAN, P. and WASSERMAN, A. I. (1976): 'Tutorial on software design techniques' (IEEE Computer Society)

GARNSEY, E. and WILKINSON, M. (1994): 'Global alliance in high technology', *Long Range Planning*, December, pp. 137–146

GARVIN, D. A. (1993): 'Building a learning organisation', *Harvard Business Review*, July–August, pp. 78–91

GARVIN, D. A. (1987): 'Competing on the eight dimensions of quality', *Harvard Business Review*, November–December, pp. 101–109

GE WEB SITE, *http://www.ge.com*

GOLDRATT, E. M. and COX, J. (1984): 'The goal—excellence in manufacturing' (North River Press)

GOMES-CASSERES, B. (1994): 'Group versus group: how alliance networks compete', *Harvard Business Review*, July–August, pp. 62–74

HAMBRECHT and QUIST WWW SITE, *http://www.hamquist.com*

HAMMER, M. and CHAMPY, J. (1993): 'Reengineering the corporation—a manifesto for business revolution' (Harper Business)

HAYES, R. H. and PISANO, G. P. (1994): 'Beyond world-class: the new manufacturing strategy', *Harvard Business Review*, January–February, pp. 77–86

HUMBLE, J., JACKSON, D. and THOMSON, A. (1994): 'The strategic power of corporate values', *Long Range Planning*, December, pp. 28–41

IRISH, V. (1994): 'Intellectual property rights for engineers' (IEE Books, Management of Technology Series)

JOHNS, J. C. (1970): 'Design methods' (Wiley Interscience)

JOHNSON, G. and SCHOLES, K. (1993): 'Exploring corporate strategy' (Prentice Hall)

KANTER, R. M. (1989): 'When giants learn to dance' (Simon and Schuster)

KANTER, R. M. (1994): 'Collaborative advantage—the art of alliances', *Harvard Business Review*, July–August, pp. 96–108

KEOUGH, M. and DOMAN, A. (1992): 'The CEO as organization designer—an interview with Professor Jay W. Forrester, the founder of system dynamics', *McKinsey Quarterly*, (2), pp. 3–30

KIDDER, T. (1982): 'The soul of a new machine' (Avon)

KONO, T. (1994): 'Changing a company's strategy and culture', *Long Range Planning*, October, pp. 85–97

LEI, D. (1993): 'Offensive and defensive uses of alliances', *Long Range Planning*, August, pp. 32–41

LORANGE, P., ROOS, J. and BRØNN, P. S. (1992): 'Building successful strategic alliances', *Long Range Planning*, December, pp. 10–17

LORENZ, C. (1994): 'Harnessing design as a strategic resource', *Long Range Planning*, October, pp. 73–84

MADU, C. N. and KUEI, C. (1993): 'Introducing strategic quality management', *Long Range Planning*, December, pp. 121–131

MATTHYSONS, P. and VAN DEN BULTE, C. (1994): 'Getting closer and nicer: partnerships in the supply chain', *Long Range Planning*, February, pp. 72–83

MCCORMACK, M. H. (1984): 'What they don't teach you at Harvard Business School', (Collins)

MOORE, J. F. (1993): 'Predators and prey: the new ecology of competition', *Harvard Business Review*, May–June, pp. 75–86

MORRIS, C. R. and FERGUSON, C. H. (1993): 'How architecture wins technology wars', *Harvard Business Review*, March–April, pp. 86–96

MUNRO-FAURE, L. and MUNRO-FAURE, M. (1992): 'Implementing total quality management' (Longman)

NASH, T. (Ed.) (1995): 'Management buy-outs—a director's guide', (Director Publications Ltd for the Institute of Directors and Phildrew Ventures Limited)

OHINATA, Y. (1994): 'Benchmarking: the Japanese experience', *Long Range Planning*, August, pp. 48–53

PEKAR, P. Jr. and ALLIO, R. (1994): 'Making alliances work—guidelines for success', *Long Range Planning*, August, pp. 54–65

PEREDIS, T. (1993): 'Strategic alliances for smaller firms', *IEEE Engineering Management Review*, Fall, pp. 66–71

PINE, B. J. II, VICTOR, B. and BOYNTON, A. C. (1993): 'Making mass customisation work', *Harvard Business Review*, September–October, pp. 108–119

PORTER, M. E. (1990): 'The competitive advantage of nations' (The Free Press)

PORTER, M. E. (1985): 'Competitive advantage—creating and maintaining superior performance' (The Free Press)

PORTER, M. E. (1980): 'Competitive strategy—techniques for analysing industries and competitors' (The Free Press)

PROKESH, S. E. (1993): 'Managing chaos at the high-tech frontier: an interview with Silicon Graphics's Ed McCracken', *Harvard Business Review*, November–December, pp. 135–144

RIGGS, H. E. (1983): 'Managing high-technology companies', (Wadsworth)

SAFFO, P. (1995): IBM,http://www.raleigh.ibm.com/for/forum7.html

SASAKU, T. (1993): 'What the Japanese have learned from strategic alliances', *Long Range Planning*, December, pp. 41–53

SHETTY, Y. K. (1993): 'Aiming high: competitive benchmarking for superior performance', *Long Range Planning*, February, pp. 39–44

SKINNER, W. (1969): 'Manufacturing—missing link in corporate strategy', *Harvard Business Review*, May–June, pp. 136–145

SPEKMAN, R. E., KANAUFF, J. W. and SALMOND, D. J. (1994): 'At last purchasing is becoming strategic', *Long Range Planning*, April, pp. 76–84

STAFFORD, E. R. (1994): 'Using co-operative strategies to make alliances work', *Long Range Planning*, June, pp. 64–74

'The Achilles' heel of Europe', *Financial Times*, 5 March 1996

'Techinvest', available from Techinvest Ltd, 31 Upper Mount Street, Dublin 2, Ireland

TOWNER, S. J. (1994): 'Four ways to accelerate new product development', *Long Range Planning*, April, pp. 57–65

TUFANO, P. (1996): 'How financial engineering can advance corporate strategy', *Harvard Business Review*, January–February, pp. 136–146

TURPIN, D. (1993): 'Strategic alliances with Japanese firms: myths and realities', *Long Range Planning*, August, pp. 11–15

TWISS, B. C. (1988): 'Business for engineers' (IEE Books, Management of Technology Series)

VENKATESAN, R. (1992): 'Strategic sourcing—to make or not to make', *Harvard Business Review*, November–December, pp. 98–107

WALKER, M. (1993): 'Cost-effective product development', *Long Range Planning*, August, pp. 64–66

WASSERMAN, A. I. and FREEMAN, P. (1976): 'Tutorial on software design techniques', (IEEE Computer Society, Catalog No. 76CH1145-2 C)

WOMACK, J. P. and JONES, D. T. (1994): 'From lean production to the lean enterprise', *Harvard Business Review*, March–April, pp. 93–103

ZAHRA, S. A., NASH, S. and BICKFORD, D. J. (1994): 'Creating a competitive advantage from technological pioneering', *IEEE Engineering Management Review*, Spring, pp. 76–85

Index

9 780852 968918